Communication for the Workplace

An Integrated Language Approach

Third Edition

BLANCHE ETTINGER, ED. D.
EDDA PERFETTO, ED. D.

JOB SKILLS – NETEFFECT SERIES

Upper Saddle River, New Jersey 07458

Library of Congress Cataloging-in-Publication Data
Ettinger, Blanche.
Communication for the workplace : an integrated language approach / Blanche Ettinger, Edda Perfetto.—3rd ed.
p. cm.—(Job skills-NetEffect series)
ISBN 0-13-118385-0 (alk. paper)
1. Business communication. 2. Business writing. I. Perfetto, Edda. II. Title.
III. Neteffect series. Job skills.
HF5718.E88 2006
808′.06665—dc22

2005005080

Director of Production and Manufacturing: Bruce Johnson
Senior Acquisitions Editor: Gary Bauer
Editorial Assistant: Jacqueline Knapke
Editorial Assistant: Cyrenne Bolt de Freitas
Senior Marketing Manager: Leigh Ann Sims
Managing Editor—Production: Mary Carnis
Manufacturing Buyer: Ilene Sanford
Production Liaison: Denise Brown
Full-Service Production: Melissa Scott/Carlisle Publishers Services
Composition: Carlisle Communications, Ltd.
Senior Design Coordinator/Cover Design: Christopher Weigand
Cover Printer: Courier Stoughton
Printer/Binder: Courier Stoughton

This book was set in GoudyWtcTReg by Carlisle Communications, Ltd. It was printed and bound by Courier Stoughton. The cover was printed by Courier Stoughton.

Pearson Education Ltd.
Pearson Education Singapore Pte. Ltd.
Pearson Education Canada, Ltd.
Pearson Education—Japan
Pearson Education Australia Pty. Limited
Pearson Education North Asia Ltd.
Pearson Education de Mexico, S. A. de C. V.
Pearson Education Malaysia Pte. Ltd.

10 9 8 7 6 5 4 3 2 1
ISBN: 0-13-118385-0

Contents

Communication for the W

An Integra Langua

Preface

The third edition of ***Communication for the Workplace: An Integrated Language Approach*** has been updated to reflect the continuous demands of business for employees who possess a mastery of communication skills.

From the very beginning of the book, its no-frills approach helps learners get right to the core of basic problems in written communications. Students are involved in the learning process through a step-by-step approach in developing their writing skills. The text helps users progress from a novice level of writing to a stage at which they are more confident in composing. Learners are encouraged to develop their critical thinking, decision-making, and problem-solving skills. Each chapter progressively builds the writing skills through the following:

- Drill reviews of basic grammar and punctuation principles
- Application exercises relating to specific composing deficiencies
- Proofreading and editing activities
- Business-related case studies that require composing letters and memorandums
- Business report writing
- E-mail writing

This user-friendly text-workbook can be used individually, in cooperative group learning, and in traditional classroom settings. It is an excellent reference guide because it offers an abundance of information on English usage and business writing.

The goal of this book is to make each user a successful communicator in today's diverse workplace.

We are indebted to our students who have inspired us throughout our writings to search for alternative instructional strategies that encourage learning.

For our families' continued support and understanding, we extend our appreciation. We are grateful to Elizabeth Sugg for her confidence in us. We thank Marie Ruvo for her time and assistance in the preparation of the manuscript.

Special thanks to the reviewers of this text: Deborah Valentine of Emory University, GA, and Carolyn Ash of University of Houston, Downtown, TX.

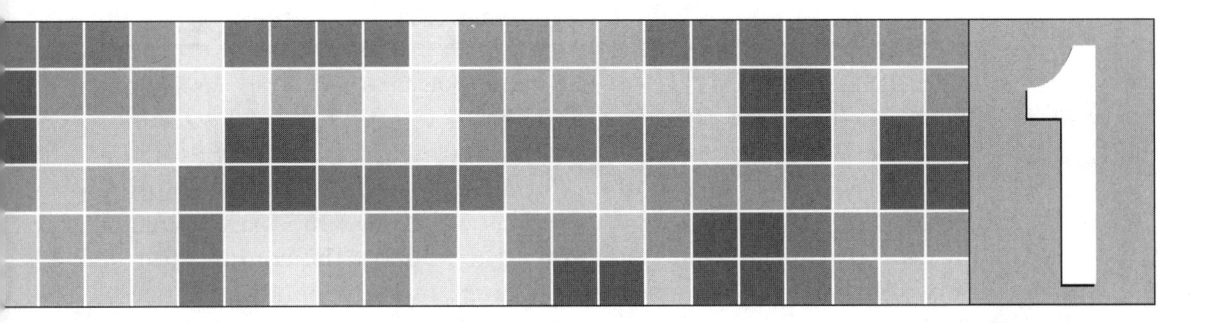

Banking on Good Sentence Structure

WHAT YOU WILL LEARN IN THIS CHAPTER

- To analyze basic sentence structure
- To apply correct capitalization
- To develop a business vocabulary in **BANKING**
- To proofread and edit correspondence and reports
- To acquire the skills for effective writing
- To compose an informational letter
- To compose promotional letters

Corporate Snapshot

WASHINGTON MUTUAL's belief in change for the better has been a strategy for success. The bank was formed to offer financial assistance to the Seattle community after the devastating Great Fire of 1889. From a small West Coast bank, the customer-oriented Washington Mutual has expanded to become one of the nation's largest mortgage lenders.

Washington Mutual is committed to support the communities it serves. It has been a leader in providing affordable lending to neighborhoods across the country. Such a program has been especially beneficial to lower-income families.

Employees are encouraged to give back to their communities. They can take up to four hours of paid time per month to volunteer their services. Washington Mutual is respected as a good employer and a good neighbor.

Source: http://www.wamu.com/home.htm

ACTIVITY A LANGUAGE ARTS RECALL

Introduction

Today's business environment requires effective communications to run a successful organization. Both written and oral communications portray a company's philosophy, policies, and attitudes. Employees must be able to communicate effectively with management, coworkers, customers, and the general public.

Years ago, business communication consisted mainly of writing letters, memos, and reports. With the increasing growth of technology, new methods of communication are used today, such as e-mail, faxes, intranet, networking, and cell phone text messaging.

Activities in this text are designed to provide you with the skills to become a proficient writer and communicator.

Sentence Structure

You must write in complete sentences to express your thoughts clearly and accurately. Sentences are varied from simple statements to more complex patterns. This chapter will review the basic format of a sentence. In succeeding chapters, more intricate patterns intended to enrich content will be reviewed.

Rule 1.1: A **sentence** is a group of words that expresses a complete thought; it consists of a subject and a verb with its modifiers (predicate), and it makes sense grammatically (independent clause).

subject verb
↑ ↑

Example: Banks offer many services to their customers.

A **subject** is a person(s) or thing(s) spoken about. The subject is usually a noun or pronoun. **Nouns** are persons, places, things, characteristics, ideas, or activities. **Pronouns** are substitutes for nouns.

Examples of nouns: employee, money, Chicago, vault, honesty, savings

Examples of pronouns: he, me, they, you

A **verb** indicates what the **subject** does or is, or what is happening to it.

Examples: author ***writes,*** employees ***are motivated,*** the amendment ***passed***

Rule 1.2: A group of words may be a sentence **fragment** if it does not express a complete thought. (Sentence fragments are grammatically incorrect.)

Example: Banks that offer many services. (Does not express a complete thought.)

Capitalization

Rule 1.3: The first word of a sentence is capitalized.

Example: ***O***ur local bank will be closed on Labor Day.

Rule 1.4: Proper names, days of the week, and months are capitalized.

Example: The ***H***artsdale ***S***avings ***B***ank will have extended hours beginning ***M***onday, ***S***eptember 1.

Rule 1.5: Abbreviations of well-known organizations and common business terms are usually capitalized.

Examples: FDIC (Federal Deposit Insurance Corporation)
ATM (automated teller machine)
IRA (individual retirement account)

Rule 1.6: Capitalize all major words in the title of complete published works, such as books, periodicals, and newspapers. Do not capitalize articles (the), conjunctions (and), or short prepositions (with) unless they begin the title. Italicize the complete title of a published work.

Examples: Our local newspaper, ***Daily Review,*** carries a weekly column on consumer saving.

The American Council for Banking Reform has published a booklet titled *The **C**onsumer's **G**uide to **M**anaging **I**ncome.*

Rule 1.7: Capitalize words derived from proper nouns.

Examples: American (derived from "America")
Japanese (derived from "Japan")

Note: These derivatives are used as adjectives, words that describe a noun.

Example: The ***Canadian*** bank will open branches in Eastern Europe.

Check for Understanding

The following exercise will help you to identify a complete sentence versus a sentence fragment.

Directions: Read each of the following statements and complete the questions that follow.

1. Community Savings Bank is a full-service bank.
2. A credit line of $2,500 from First National Bank.
3. Savings deposits are usually insured by the Federal Deposit Insurance Corporation to a maximum of $100,000.
4. The book of checks that you ordered.

Questions:

a. What do statements 1 and 3 have in common relative to their structure?

____

b. How are statements 2 and 4 different from 1 and 3 in their structure?

____

c. What part of speech is missing to complete statements 2 and 4?

____

(Answers are given below.)

Answers to "Check for Understanding"

a. complete sentence b. sentence fragment c. verb

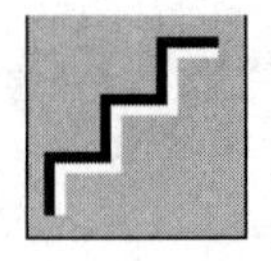

Building English Power: Sentence Structure

Answers to "Building English Power" in each chapter can be found in Appendix A. Complete this activity before checking answers.

Exercise 1. In the blanks provided on the left, indicate whether the sentences are complete (C) or incomplete (I). In the complete sentences, underline the subject with a single line and the verb with a double line.

__________ 1. Your monthly itemized billing statement lists.

__________ 2. We invite you to apply for our special credit card.

__________ 3. Finance charges do not appear on this month's statement.

__________ 4. Free additional credit cards for your family.

__________ 5. No annual fee for the first year.

__________ 6. To access an ATM (automated teller machine), you need the customer's personal identification number.

__________ 7. Private loan companies that offer personal loans and mortgages.

__________ 8. As a valued customer, you have a minimum credit line of $5,000.

____________ 9. The bank requested collateral before issuing the loan.

____________ 10. If you pay your credit card balance in full each month to avoid interest charges.

____________ 11. The financial advisor at our local bank was helpful in suggesting several investment options.

____________ 12. Direct deposit is encouraged.

____________ 13. At the end of the calendar year, the bank will inform you of the total interest accrued on all accounts.

____________ 14. Tracing a lost check.

____________ 15. Banks require that you provide identification when cashing a check.

Exercise 2. Rewrite each incomplete sentence which you identified in Exercise 1 to make it a complete sentence. Add words if necessary.

Note: Writing tasks may be completed on a computer.

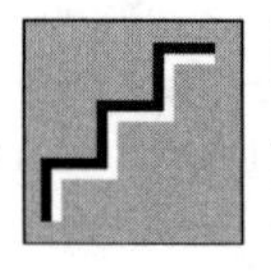

Building English Power: Capitalization

Directions: In the following sentences, circle each lowercase letter that should be capitalized. Indicate italics by underlining. Corresponding rules follow each sentence.

1. you must reply by september 1 to receive a free desk calendar. (1.3, 1.4)

2. mastercard and visa are the most popular bank credit cards. (1.3, 1.4)
3. new york bank will be closed on saturdays during july and august. (1.3, 1.4)
4. most savings accounts are insured by the fdic, a government agency known as the federal deposit insurance corporation. (1.3, 1.5)
5. what did you think of mr. baker's speech in san francisco on banking reforms? (1.3, 1.4)
6. the federal home loan mortgage corporation announced that mortgage rates are on the rise. (1.3)
7. eastern trust offers an ira that has no annual fee. (1.3, 1.4)
8. a cd account offers the investor a higher rate of interest. (1.3, 1.5)
9. the new computer software, known as bank express, will allow users to keep personal and financial records up to date. (1.3, 1.4)
10. on thursday, december 1, the royal bank will introduce its new service for investors. (1.3, 1.4)
11. dr. alan smith will speak on global investments at the orlando convention center in florida on Saturday, july 8. (1.4)
12. atms are located in many shopping malls in the united states. (1.3, 1.4, 1.5)
13. the american bank will sponsor the next world cup usa. (1.3, 1.4)
14. pathways to financial security made the new york times best-seller booklist. (1.3, 1.4, 1.6)
15. the european bank will expand its offices to singapore and hong kong. (1.3, 1.4, 1.7)

ACTIVITY B BUILDING WORD POWER

A rich vocabulary gives an individual superior communication skills. Such an individual is a valuable asset to a company that strives to portray a positive image in the marketplace.

A well-written letter or report reflects the strong use of words. In this activity throughout the text, you will be able to increase your vocabulary of particular industries.

Directions: The words below are commonly used in the banking industry. From this list, select the word or phrase that correctly completes each statement. Before completing the vocabulary exercise in each chapter, refer to the Personal Dictionary in Appendix F for the meaning of each word.

accrued	FDIC
ATM	global
authorization	insufficient funds
electronic banking	IRAs
CDs	mortgage
co-applicant	network
collateral	specialists
debit card	transactions

__________ 1. The depositor had ________ in the account to pay the bill.

__________ 2. With ________, customers can use computers, telephones, or teller machines to conduct banking business.

__________ 3. To meet ________ competition, business is required to understand the needs of foreign markets.

__________ 4. A(n) ________ is a long-term loan obtained by an individual to purchase a home; the debt is repaid to the bank on a monthly basis.

__________ 5. The client gave written ________ to his lawyer to sell his property.

__________ 6. Banking ________ frequently involve the transfer of money.

__________ 7. ________ are experts who are consulted in specific areas, such as home mortgages, foreign investments, and bonds.

__________ 8. How much interest has ________ on my CD account?

__________ 9. His sister was willing to be a(n) ________ in assuming the financial responsibility for repayment of the loan.

__________ 10. The firm's computer ________ links the main office with its branches throughout the country.

__________ 11. Banks offer specialized savings accounts, known as ________, that cannot be withdrawn without a penalty until the depositor reaches retirement age.

_____________ 12. The greatest advantage of a(n) ________ is the opportunity to withdraw cash 24 hours a day without interacting with a bank employee.

_____________ 13. Individuals can pledge as ________ their automobiles, jewelry, bank accounts, and stocks to guarantee a loan.

_____________ 14. The government agency, the ________, insures savings accounts to a maximum of $100,000.

_____________ 15. Savings banks offer ________ for one, two, and three years. Although these investments generally offer a higher interest rate, they carry a penalty for early withdrawal.

_____________ 16. I use my ________ when paying for the grocery bill at the supermarket.

(Answers to "Building Word Power" in each chapter can be found in Appendix A.)

On the Alert

- A sentence that makes a statement or gives information is a *declarative sentence* and *ends with a period.*

  Example: U.S. banks invest in foreign countries.

- A sentence that begins with the words *who, what, where, when, why,* or *how* is a question and ends with a question mark (?).

  Example: Who is the new bank manager?

- Common business abbreviations are acceptable when understood by the reader. Plurals of abbreviations that are two or more letters are formed without an apostrophe.

  Examples: ATM, IRA, ATMs, IRAs

- Periods are generally omitted in abbreviations that are all capitalized.

  Examples: FDIC, CD (certificate of deposit)

- access (noun)—means of entry, approach

  Example: John had ***access*** to the bank vault.

- excess (noun)—surplus, extra

  Example: The ***excess*** will be distributed to the stockholders.

Editing and Proofreading

After the initial draft of a document is completed, editing and proofreading become the next stage in the writing process. The completed document should meet the criteria of the six Cs: clear, complete, concise, consistent, correct, and courteous. Editing is the final review of the main idea. The message must be checked to ensure it is effective in its structure and readability. Changes should then be made to enhance the writing.

Proofreading focuses on correcting mechanical errors in spelling, punctuation, and grammar.

Checklist for Effective Writing

___ Content focused on main idea of the document

___ Correct sentence structure; no run-on or incomplete sentences

___ Even flow of sentences for smooth reading

___ Varied use of words rather than repetitive use; ex. I

___ Correct spelling, grammar, punctuation

Activities C and D in each chapter will help you refine your important skills in writing.

ACTIVITY C INDUSTRY PERSPECTIVES: BANKING TRENDS

In this section of each chapter, new trends of a particular industry are discussed. "Industry Perspectives" is presented as a report that requires you to proofread for English usage, spelling, punctuation, and typographical errors.

Directions: Using appropriate revision marks, proofread, correct, and key the following report. (See Appendix E for explanation of proofreaders' marks.)

Rule Reference	
	Electronic banking service has become very popular in
1.4	banking transactions in america. This is also true in
1.7	european and asian countries. You may us an automated

teller machine to make deposits or withdrawals without
1.3 interacting with a bank employee. in automated systems,
you may withdraw cash immediately from your checking or
savings account. Deposits can also be made to these
accounts, and funds between accounts can be exchanged
easily. You simply insert a card with your personal
1.5, 1.2 identification code number at at an atm station. Which
is available at most banks.

1.2 You can do your banking. In any part of the country
and at any time through a computer networking system.
Electronic banking service is expanding to include direct
payroll deposits, retail store purchase transactions,
and personal bankking transactions from a home computer.

1.3 for individuals with busy schedules, electronic banking
1.4 is available seven days a week, monday through sunday.

Did you find these errors?
8 capitalization errors
2 sentence fragments
3 typographical/spelling

See Appendix B for models of business and personal letters.

ACTIVITY D EDITING POWER

When editing correspondence, you should read carefully for correct English usage, punctuation, content, and clarity. This exercise in "Editing Power" will help you refine your ability to write logically and accurately.

Task A: Edit, format, and proofread the following letter for errors. Use proofreaders' symbols to mark the errors; then key the letter correctly.

Errors
- Sentence fragments
- Capitalization
- Typographical
- Italicize published work

Date

MR JOSEPH ROBINSON
131 POST ROAD
DULUTH MN 55803-5122

Dear Mr. Robinson

Thank you for choosing fidelity savings bank for your banking needs. We are confident you will be happy doing business with us.

we are enclosing our latet brochure, banking made easy for the consumer. Which describes teh many services we offer customers. These services include free checking with interest, direct deposits, iras, cds, and life insurance. Our bank is open six days a week monday to friday, from 9 a.m. to 3 p.m. and on saturday until noon. For your convenience, we have evening banking hours on tuesday and thursday. We are also expanding our canadian services by opening branches on october 10 in montreal and quebec. You will benefit from the many services offered by fidelity savings bank, a member of the federal deposit insurance corporation.

Cordially yours,

Peter Gerena
Manager
ri
Enclosure

Task B: Edit, format, and proofread the following unarranged letter for all errors, then key it correctly. (See Appendix B for model of letter format.)

INTERNATIONAL ELECTRONICS COMPANY, 14 CENTRAL AVENUE, INDIANAPOLIS IN 46204-1352 Ladies and Gentlemen: american trust company is happy to announce the formation of a of a new department to serve ourr international clients. we will be able to serve corporations and financial institutions with branch offices in europe and asia. Global Services, the name of this department, will be staffed by specialists who understaand the foreign market and can supply you with information to help you make profitable business decisions.

You may contact us at our main office on 56 river road to open an account. So that we may assist your firm in meeting international competition. Yours truly, Frank Jarcho, Vice President.

ACTIVITY E WRITING PRACTICUM

This "Writing Practicum" section will help you to improve your written communication skills. Exercise(s) in each chapter will focus on writing problems people frequently encounter. By concentrating on specific problem areas in the writing process, you will build a strong foundation. This will result in writing meaningful and effective business communications.

A common writing problem is sentence fragments. As you learned, a **sentence fragment** is an incomplete thought. The focus of this writing practicum is to recognize sentence fragments and to revise each one into a complete sentence that has a subject and a verb and that expresses a complete thought.

EXERCISE 1 WRITING IN COMPLETE THOUGHTS

Either begin or end each sentence with a complete thought. Use the blank space provided to complete each sentence.

1. When a small Midwestern bank failed, the FDIC assured depositors ________________.
2. ________________ is one of the features offered by a full-service bank.
3. Mr. Sanders received a free gift from his bank when ________________.
4. One of the restrictions in a CD account is ________________.
5. As interest rates dropped, many customers of the bank chose ________________.
6. ________________ is one of the ways in which banks make money.
7. The slow teller at the second window was much less efficient than ________________.
8. If mortgage rates continue to increase, ________________.
9. The letter from Mr. Robinson, the bank president, explained ________________.
10. ________________ is one of the advantages of a checking account.

(Answers to "Writing Practicum" in each chapter are in Appendix A.)

EXERCISE 2 COMBINING SENTENCES

Sentences should not be wordy or choppy but should flow for clarity and easy reading. Your goal is to combine each set of sentences into one meaningful sentence that expresses the main idea. You may add or delete words.

Example

National Trust is a bank.
It has opened a branch office in our neighborhood.
It has Saturday banking hours.

Combined Sentence: National Trust Bank opened a branch office in our neighborhood that has Saturday banking hours.

1. ATMs are convenient.
   Busy persons may conduct banking transactions 24 hours a day.
   ATMs are open 24 hours a day.
   Combined Sentence: ______________________________
2. The bank has a special rate on loans.
   The loan rate for a car is 7 percent.
   It must be an American-made car.
   Combined Sentence: ______________________________
3. A certificate of deposit is a term savings account.
   It pays a higher rate of interest.
   It carries a penalty for early withdrawal of money.
   Combined Sentence: ______________________________
4. The Internet is used for banking transactions.
   Many depositors have online accounts.
   Depositors can download statements and pay their bills.
   Combined Sentence: ______________________________
5. The bank offers a basic checking account.
   With a basic checking account, you can open the account for as little as $25.
   With a basic checking account, you do not need to maintain a minimum balance.
   Combined Sentence: ______________________________

EXERCISE 3 DEVELOPING MEANINGFUL PARAGRAPHS

The following passage needs to be rewritten for correctness of expression and logical flow of thoughts.

You may change any sentence that you believe is incorrect. Examine each sentence and decide whether you agree to keep the original sentence or rewrite it. In the blank space provided after each sentence, indicate whether you will keep (K) or rewrite (R) the sentence.

Rearrange the order of the sentences so that the ideas are related and there is a logical flow of thought forming meaningful paragraphs, then key the passage.

1. Banking in America has changed dramatically over the past decade. ________
2. Banks have responded to changing needs of the consumer in the local communities. ________
3. The electronic age has introduced ATMs, direct deposits, and computer transactions, which improved banking services. ________
4. Customers now receive fast, without long waiting lines, efficient service. ________
5. Also, with electronic banking systems, customers can transact their financial business from any area of the country. ________
6. Also, a large number of banks have merged to offer additional services that facilitate handling global transactions. ________
7. They are open some evenings, Saturdays, and Sunday mornings. ________
8. Banks are prepared to meet the challenges of the next millennium with these changes. ________

- List in order of preference the sentence numbers, then key the passage.

______ ______ ______ ______ ______ ______ ______ ______

EXERCISE 4 WRITING OPENING AND CLOSING SENTENCES IN LETTERS

From each group choose the best opening and closing sentences that would appear in letters from a bank. Write your answer, A or B, in the space provided.

Opening Sentences

It is important that the tone and clarity of the message be set at the beginning of the correspondence. In a positive manner, attract the reader's intention to the message.

______ 1. a. Some banks charge fees. Does yours?
b. Are you paying a fee for your bank credit card?

______ 2. a. We want you as a Webb customer, and to prove it we're offering you many free banking services.
b. At Webb we are offering many free banking services.

______ 3. a. North Branch Bank is expanding into Westchester County.
b. North Branch Bank, with more than $10 billion in assets, is now building branches to serve Westchester County.

______ 4. a. The Money Master Debit Card from Suburban Bank is convenient and fast.
b. At Suburban Bank, we understand your busy life and are offering the Money Master Debit Card.

Closing Sentences

Closings should end with a courteous, positive statement. Also, offer suggested action to be taken.

______ 5. a. Feel free to talk to our bank counselors at 1-800-465-2000.
b. Our bank counselors are available to help you. We invite you to call us at 1-800-465-2000 or visit us at any of our branches.

______ 6. a. Call now and protect your future.
b. Call now to protect your future; open an IRA with us.

______ 7. a. We invite you to take advantage of this special offer.
b. Come in and take advantage of this special offer.

Good Letter Writing

Effective business correspondence is essential for an organization to maintain a positive image with customers and to obtain the desired results. Business letters are used to communicate formal messages to persons outside the organization. A good communicator must use the six Cs of writing. The correspondence should be *clear, complete, concise, correct, consistent,* and *courteous.* An effective business letter includes the following:

- State purpose of letter (to inform, to respond, to request, to persuade, to express a feeling).
- Focus on a specific audience (global, local, or internal audience).
- Gather accurate information.
- Organize information (opening, middle, closing).
- Use correct format and attractive appearance.
- Communicate a friendly tone.
- Use courteous closing.

In each chapter, "Writing Pointers" highlight essential items that need to be considered before writing a specific type of business correspondence effectively. These pointers give helpful suggestions to aid you in composing your response to the case studies in Activity F. Study "Writing Pointers" before composing the correspondence.

The "Model Letters" that appear in each chapter offer another opportunity to review good letter writing. The models refer to specific types of business letters discussed in the chapter.

Writing Pointers

Informational Letter

- Focus on the information to be disseminated.
- Write to the point.
- State main idea immediately.
- Give specific facts and details.
- End with a courteous, positive tone.

Promotional Letter

- Attract attention with a strong opening statement.
- Build up the reader's curiosity to react to the promotion.
- State the objectives of the promotional offer.
- Appeal to reader's needs.
- Be specific when indicating benefits.
- Use appealing language.
- Close with action statement.

A form letter is used to:

- Reply to frequent requests for information.
- Communicate information to a large number of people.

Model Informational Letter

Date

MR AND MRS JAMES STEVENS
14 MAPLE DRIVE
DOBBS FERRY NY 10532-1600

Dear Mr. and Mrs. Stevens:

Opening focuses on message

Appeals to specific reader's needs

Today's high cost of a college education often poses a financial hardship on the family. County Trust Bank is offering a special educational loan to its depositors.

Gives specific facts

For a limited time only, you will be able to obtain a loan at a special rate of 5 percent per annum. A loan kit explains in detail the specific benefits and affordable options. Fill out the enclosed card for your complimentary copy or call us at 1-800-623-5000.

Positive ending

If you wish to discuss this loan offer with one of our representatives, stop in at any one of our branch offices. We will be happy to help you get a loan.

Sincerely yours,

Ruth Marlowe
Financial Officer
ri
Enclosure

Model Promotional Letter (form letter)

Date

Dear Depositor:

Question attracts attention

Are you facing the dilemma of the high cost of sending your child to college?

Appeals to specific reader's needs
Gives specific facts

County Trust Bank is introducing a new educational loan plan to help you meet the rising cost of a college education. Our plan is geared to families whose income falls below $50,000, and the introductory rate for the loan is only 5 percent. If you act immediately, you will receive a loan kit that explains in detail the specific benefits and affordable options.

Positive ending

Our representatives will gladly help you to apply for this outstanding offer. Call us at 1-800-623-5000 and start saving money immediately.

Sincerely yours,

Ruth Marlowe
Financial Officer

ri

ACTIVITY F WRITING POWER: COMPOSING CORRESPONDENCE

Now that you have reviewed the elements of English, expanded your business terminology, developed your editing and proofreading skills, and sharpened your ability to compose, you are ready to prepare correspondence based on practical business and personal business situations.

In each chapter, Activity F presents case studies. In Chapters 1, 2, and 3, Case Study 1 is given as an exercise to help you to organize your ideas in composing a letter that is simple, logical, and to the point.

CASE STUDY 1: PROMOTIONAL LETTER

Read Case Study 1. Then read Options A, B, and C for each section of the letter. Select the option that you believe is most appropriate for the opening, body, and closing statements. Then key the entire letter based on the options you selected.

Case: You were asked by your bank manager, Mr. Eric Sanders, to draft a promotional letter to inform customers of a new premium credit card. Key the letter as a form letter. Do not include an inside address. Use "Dear Customer:" for the salutation.

Opening sentence:

a. As a customer, you have the opportunity of getting a premium credit card from us.

b. As our customer, we invite you to consider choosing our new premium card.

c. As one of our valued customers, we are offering you the opportunity to choose our premium credit card.

Option A, B, or C _____

Sentences in body of letter

Sentence 1:

a. The card provides superior benefits and services without an annual fee.

b. Our card provides superior benefits and service. There is no annual fee.

c. The card gives you, the cardholder, benefits and services. No annual fee is charged.

Option A, B, or C _____

Sentence 2:

a. If you have a monthly balance, you will be charged a rate that is below the average of other financial institutions.

b. Our finance charge for unpaid balances is below the average rate required by other financial institutions.

c. Our charge for unpaid balances is below the average rate charged by other financial institutions.

Option A, B, or C _____

Sentence 3:

a. The card also offers a 90-day protection on purchases against theft or damage.

b. It also offers 90 days of protection on purchases against theft or damage.

c. It offers 90-day protection against theft or damage of merchandise.

Option A, B, or C _____

Sentence 4:

a. If you decide you want more cash, you can get the cash at an ATM.

b. You will be able to get cash at any ATM when you need it.

c. ATMs offer cash.

Option A, B, or C _____

Closing statement:

a. Call our toll-free customer service for your valued card.

b. We hope to hear from you soon.

c. There is 24-hour customer service at 1-800-255-5000. Complete the enclosed application form or call one of our representatives to receive your valued card.

Option A, B, or C _____

Note: Appendix B includes model formats of business correspondence.

CASE STUDY 2: INFORMATIONAL LETTER

In each chapter, read Case Study 2, then respond to the questions that follow to help you organize your ideas before beginning to compose the correspondence. After you have completed the questions, compose the correspondence. **Do not copy sentences as presented in any of**

the case studies. Remember when composing correspondence in any chapter use the correct format. Be creative and remember to add a closing line to retain the goodwill of the customer.

Case: Customers have been complaining that the banking hours are inadequate. Therefore, the County Trust Bank is adding hours on Thursday until 6 p.m. and on Saturdays from 9 a.m. until 1 p.m. Full banking services will be available to customers in the loan, investment, and insurance departments. Draft a form letter to bank customers notifying them of this added service. Use "**Dear Depositor:**" for the salutation.

Organize Your Ideas

1. What is the most important point you want to make to the bank's customers?

____

____

____

2. What should you include in the opening statement?

____

____

____

3. What specific details should you include in the body?

____

____

____

4. What should you include in your closing statement to demonstrate the bank's goodwill toward its customers?

____

____

____

CASE STUDY 3: PROMOTIONAL FORM LETTER

In each chapter, read Case Study 3, organize your ideas, and compose the correspondence.

> Note: The following case should be written as a form letter that is used when the same message is sent to many addresses. Inside addresses may be merged to personalize a form letter. (See letter format in the Appendix.)

Case: Draft a promotional form letter that will attract new depositors to Union Bank. Two types of CDs are being offered by the bank. One is a 12-month CD, minimum deposit $500, paying 2¼ percent. A two-year CD with a minimum of $1,000 is offered at 3 percent.

In addition, there are no fees for use of the ATM card. Also, for new depositors the bank is offering a free consultation with the bank's financial advisor. Encourage readers to open an account soon so they can take advantage of this limited offer. The letter should be signed by Mario Lopez, Bank Manager.

CASE STUDY 4: PROMOTIONAL LETTER

To challenge you further, Case Study 4 in each chapter will present a case for which you will prepare the correspondence.

Case: The Empire State Bank wants to announce a limited-time offer for automobile loans at a low interest rate of 7½ percent. This loan will be available only for the month of May. Ms. Laureen Ramos, your manager, has asked you to draft a form letter to be mailed to bank customers encouraging them to consider applying for such a loan. Individuals who want more information may contact their local branch office or call the toll-free number, 1-800-433-5000. (See letter format in Appendix B.)

ASSESSMENT: CHAPTER 1

English Power

Directions: *In the following sentences, insert correct capitalization where needed.*

1. in july, the european and american bankers will hold an international meeting in london.
2. the bank of seattle is offering a new ira plan to its clients.
3. the tampa tribune reported that the banks in america have raised their mortgage rates.
4. the world bank will assist the financial institutions in the emerging nations of hungary, poland, and south africa.
5. In the united states, people's banking habits have been affected by the use of atms.

Directions: *In the blanks provided, indicate whether the sentences are complete (C) or incomplete (I). In the complete sentences, underline the subject with a single line and the verb with a double line. Rewrite each incomplete sentence to make it complete and identify the subject and verb.*

______ 6. The savings bank plans to offer its one-year CD at a higher rate.

____

______ 7. Americans are faced with foreign money exchange rates while traveling in Europe and the Far East.

____

______ 8. The bank client was notified.

____

______ 9. One of the advantages of electronic banking.

____

______ 10. Our firm has contracted with People Bank.

____

Editing Power

Directions: *Edit the following form letter for errors, add words or transpose phrases where applicable, and key it correctly.*

Dear Depositor:

Thank you for choosing american Bank for your financial business.

we are pleased to offer you, a valued customer, our Classic depositors Program. This special program combines full-service cheking account with credit card privileges. There is no annual fee for your card. you have complete protection. if your card is lost or stolen. The extra benefits include a credit line of $2,000 and $100,000 travel accident insurance policy. Our many services that are available to you in the enclosed brochure.

We urge you to return the application form as soon as possible. We look forward to srving you.

Sincerely yours,

James Harley
Customer Services

ri

Enclosure

Writing Power

Directions: *After reading the case below, draft a letter. Use "Dear Depositor:" for the salutation.*

Compose a promotional form letter to be sent to the depositors of the American Bank informing them of the new Money Debit Card. This card can be used as a debit card to purchase merchandise. Explain in your letter that the debit card is a convenient way to charge purchases directly to the checking account. It is also safe since it is not necessary to carry cash. This card can also be used as an ATM card to withdraw money. There are no fees.

Whenever the card is used, the cardholder earns air travel miles that can be used to travel in the United States.

Encourage the reader to take advantage of this new card. The letter should be signed by Rose Hare, Bank Manager.

"Leadership and learning are indispensable to each other."

—*John F. Kennedy*

2

Ensuring Subject–Verb Agreement

WHAT YOU WILL LEARN IN THIS CHAPTER

- To apply correct subject–verb agreement principles in business correspondence
- To develop a business vocabulary in **INSURANCE**
- To proofread and edit correspondence and reports
- To compose a sales letter
- To compose a letter of notification

Corporate Snapshot

AFLAC is a leading provider of health, life, and supplementary insurance. Overseas, the company is the industry leader in Japan; and through advertising, the "AFLAC Duck" has boosted the company's name recognition to the general public. AFLAC's insurance business is primarily through direct payroll deductions of workers at participating companies.

The company philosophy holds that diverse employee backgrounds, experiences, and talents will ensure better customer service. AFLAC has consistently been named to *Fortune* Magazine's list of the "100 Best Companies to Work for in America."

The firm encourages its employees to volunteer their services to local organizations, allowing free time during the workday. The firm's major community project is to sponsor one of the largest pediatric cancer centers in the United States. AFLAC truly demonstrates corporate partnership with the community.

Source: *http://www.aflacny.com*

ACTIVITY A LANGUAGE ARTS RECALL

Subject–Verb Agreement

A verb must agree with its subject in number and person.

A **subject** is a noun or pronoun that identifies what is being discussed. The subject is a person, a place, a thing, or an idea.

A **verb** expresses action or a state of being. A verb tells what the subject is *doing* or how the subject is *feeling*.

Rule 2.1: A singular subject takes a singular verb; a plural subject takes a plural verb.

Examples: The ***agent is*** planning a seminar on life insurance. (singular subject, singular verb)

The two insurance ***plans offer*** similar benefits. (plural subject, plural verb)

Note: An "s" after a regular verb indicates a singular number.

Example: An employee request*s*. (singular)

The employee*s* request. (plural)

Rule 2.2: A subject that has two nouns joined by "and" takes a plural verb. (compound subject)

Example: ***Kevin and Maria are submitting*** a claims form for their property damage. (compound subject, plural verb)

Rule 2.3: A collective noun (a group of individuals acting as a whole) takes a singular verb.

Example: The ***committee is*** revising automobile insurance policies.

Rule 2.4: The pronoun "you" takes the same verb as a plural subject.

Example: ***You are*** eligible for a discount on your homeowner's policy.

Check for Understanding

The following exercise will help you to better understand subject–verb agreement.

Directions: In the spaces provided, indicate the subject and verb in each of the following sentences. Also identify whether each is singular (S) or plural (P).

1. Our policy offers a comprehensive health plan.
2. The agent and his client have agreed on the terms of the insurance policy.
3. Your insurance plan requires a second opinion for surgery coverage.
4. Management expects to offer dental insurance next month.
5. Auto insurance premiums increase with a rise in the number of accidents.
6. The review committee authorizes any change in insurance rates.

7. You always receive a rebate on your insurance premium if you enroll in a defensive driving course.
8. The insurance company and the policyholder have agreed on the amount of financial settlement.

	Subject	**Verb**	**S or P**
1.			
2.			
3.			
4.			
5.			
6.			
7.			
8.			

Question: Summarize the subject–verb rule that applies to your answers in the above exercise. ______________________________

(Answers are given below.)

Answers to "Check for Understanding"

1. policy, offers, S
2. agent and client, have agreed, P
3. plan, requires, S
4. management, expects, S
5. premiums, increase, P
6. committee, authorizes, S
7. you, receive, S
8. company and policyholder, have agreed, P

Rule: A verb must agree with its subject in number and person.

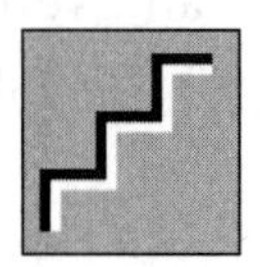

Building English Power: Subject–Verb Agreement

Directions: In each of the sentences, circle the subject; underline the correct form of the verb that agrees with the subject. (Check your answers in Appendix A.)

1. The insurance plan (protect, protects) you against personal damages. (2.1)
2. These cost-cutting procedures (help, helps) to lower premiums. (2.1)
3. Your automobile club (has, have) been able to increase services without additional costs. (2.1)
4. You (ask, asks) your insurance agent about the new 24-hour claim service. (2.4)

5. Union and management (has, have) agreed on cost-cutting procedures to lower the company's insurance expenses. (2.2)
6. After a natural disaster, the insurance industry often (face, faces) an excessive number of property damage claims. (2.3)
7. A $1-million umbrella insurance policy (was, were) available to the policyholders. (2.1)
8. The car rental contract (include, includes) property damage insurance. (2.1)
9. Michael and his friend (is, are) paying high premiums for their automobile insurance. (2.2)
10. The safe-driving course usually (qualify, qualifies) the driver for a decrease in auto insurance. (2.1)
11. Under the contract, you (become, becomes) eligible for benefits after three months of employment. (2.4)
12. Flood damage and theft (is, are) not covered under your basic automobile insurance plan. (2.2)
13. Our records (indicate, indicates) that the above balance is outstanding. (2.1)
14. The employees' negotiating team (has, have) presented management with a comprehensive health plan. (2.3)
15. Dental insurance and optical coverage (is, are) fringe benefits offered to employees by some companies. (2.2)

ACTIVITY B BUILDING WORD POWER

Directions: The words below are commonly used in the insurance industry. From this list, select the word or phrase that correctly completes each statement. Refer to the Personal Dictionary if you are uncertain of the meaning of a word. (Check your answers in Appendix A.)

accelerate, accelerated
alternative
bodily injury insurance
cafeteria-type medical insurance
catastrophic insurance
deductible
escalate, escalating
liability insurance
option
phasing out
premium
property damage insurance
subsidized
umbrella insurance

_______________ 1. To protect yourself against excessive financial claims or expenses, ________ is available to you. A(n) ________ policy covers you above the basic insurance terms.

_______________ 2. Because of a reorganization, the XYZ Company needs to eliminate departments; therefore, it will be ________ its claims department.

_______________ 3. In an attempt to reduce the ________ number of automobile accidents, some insurance companies are giving a discount for safe driving records.

_______________ 4. ________, which is very much like a restaurant menu, gives employees the ________ of selecting insurance benefits that meet their needs.

_______________ 5. The driver's ________ covered the medical costs of the injured pedestrian who was hit by the car.

_______________ 6. I will sue John Smith under the ________ clause for the damages that occurred to my car when his van hit it.

_______________ 7. Most insurance policies have a ________ clause that requires the insured to pay a minimum amount when claims are settled.

_______________ 8. You will notice a higher ________ for your insurance due to a rate increase of 10 percent.

_______________ 9. When a company cafeteria is ________, employees pay less for their meals.

_______________ 10. To pass a truck on the hill, the motorist ________ his speed.

_______________ 11. The judge gave the young offender the option of performing community service as a(n) ________ to paying a fine.

_______________ 12. ________ protects the insured against injury or property damages suffered by another person.

On the Alert

- their (possessive)—belonging to them

  Example: We mailed ***their*** insurance policy yesterday.

- there (adverb)—in or at that place

  Example: ***There*** are several changes in the policy.

- titles—Italicize the title of a complete published work, such as a book, periodical, newspaper, pamphlet, or brochure. (Do not capitalize articles, conjunctions, or short prepositions [the, and, with] unless they are the first word of a title.)

  Example: The ABCs of Auto Insurance

- million—Amounts of money in millions are expressed partially in numbers. When using the $ sign, do not write out the word *dollar* at the end of the phrase.

  Example: $1 million

ACTIVITY C INDUSTRY PERSPECTIVES: INSURANCE PROGRAMS

Directions: Using appropriate revision marks, proofread and key the following report.

Rule Reference

1.4, 2.2 In the united states, corporations and individual households is concerned with increasing costs of insurance. Companies that are faced with escalating expenses are reexamining employee insurance benefits.
2.1 One alternative is to offer cafeteria-type medical insurance plans to employees. Under these plans,
1.3 employees has the option of choosing the best medical coverage for there needs. in one firm, vacation days can be earned by those employees who select a plan that does not include every type of coverage or that has a higher deductable. The trend is away from a uniform medical plan for all of the company employees.

2.1 For retirees, health costs has accelerated so much that some companies are phasing out health insurance subsidies for future employees. In some instances, a

2.1 special stock ownership plan for employees have been

introduced. The acumulated value of the stock can be

used to pay for medical costs at some later date.

2.3 Business find these changes to be financialy

beneficial; however, employees may be affected in their

personal insurance protection. Experts in teh insurance

2.1, 2.4 field recommends that you reviews the type of insurance

that meets your personal needs.

Did you find these errors?
7 subject–verb agreement
1 confusing word
4 typographical/spelling errors
3 capitalization problems

ACTIVITY D EDITING POWER

Task A: Edit, format, and proofread the following letter. Use proofreaders' symbols to mark the errors; then key the letter correctly.

Errors
- Subject–verb agreement
- Sentence fragments
- Spelling
- Capitalization
- Typographical

MR WILLIAM SIMMS

10 MAPLE AVENUE

TOLEDO OH 43692-9987

Dear Mr. Simms:

We are happy to enclose a renewal for your automobile

insurance policy for the coming year. Please review you

policy carefully so that you are fully protected.

Our brochure, The ABCs of auto insurance, is enclosed. To help you determine whether your present policy is meeting your needs. This brochure list in detail the features we offer automobile owners. Our policy offer discounts to drivers with good safety records. For full protection, you should have adequate bodily injury and property damage coverage.

If you agree with the limits of liability and coverge included in the enclosed policy, detach the premium statement and mail it with your payment. If you have any questions, our representative are available 24 hours a day at (800) 841-9000. After doing business with us, you will agree that our reliability and customer services is superior to that of our competitors.

Our firm value policyholders like you, and we sincerly appreciate the opportunity to serve you.

Sincerely yours,

David Knight

Vice President
ri
Enclosures

Task B: Edit, format, and proofread the following unarranged letter for all errors; then key it correctly.

```
Ms Sandra Davis, 32 Leeds Place, Boston MA 02155-2199

Dear Ms. Davis:

Our company invite you, a valued customer, to consider
our Plus Insurance Policy.

This umbrella policy insure you to a maxmum of $1 million
dollars against catastrophic financail loses from
unexpected lawsuits. The cost of this umbrela insurance
policy are only a few cent a day. You and your family
enjoys peace of mind knowing that your assets are
protected.

Our expereinced representatives are available 24 hours a
day to answer any questions you may have. Sincerely
yours, William Snyder, Vice President.
```

ACTIVITY E WRITING PRACTICUM

EXERCISE 1 WRITING OPENING AND CLOSING SENTENCES

From each group, choose the best opening and closing sentences that were taken from letters sent to an insurance company. Be aware that

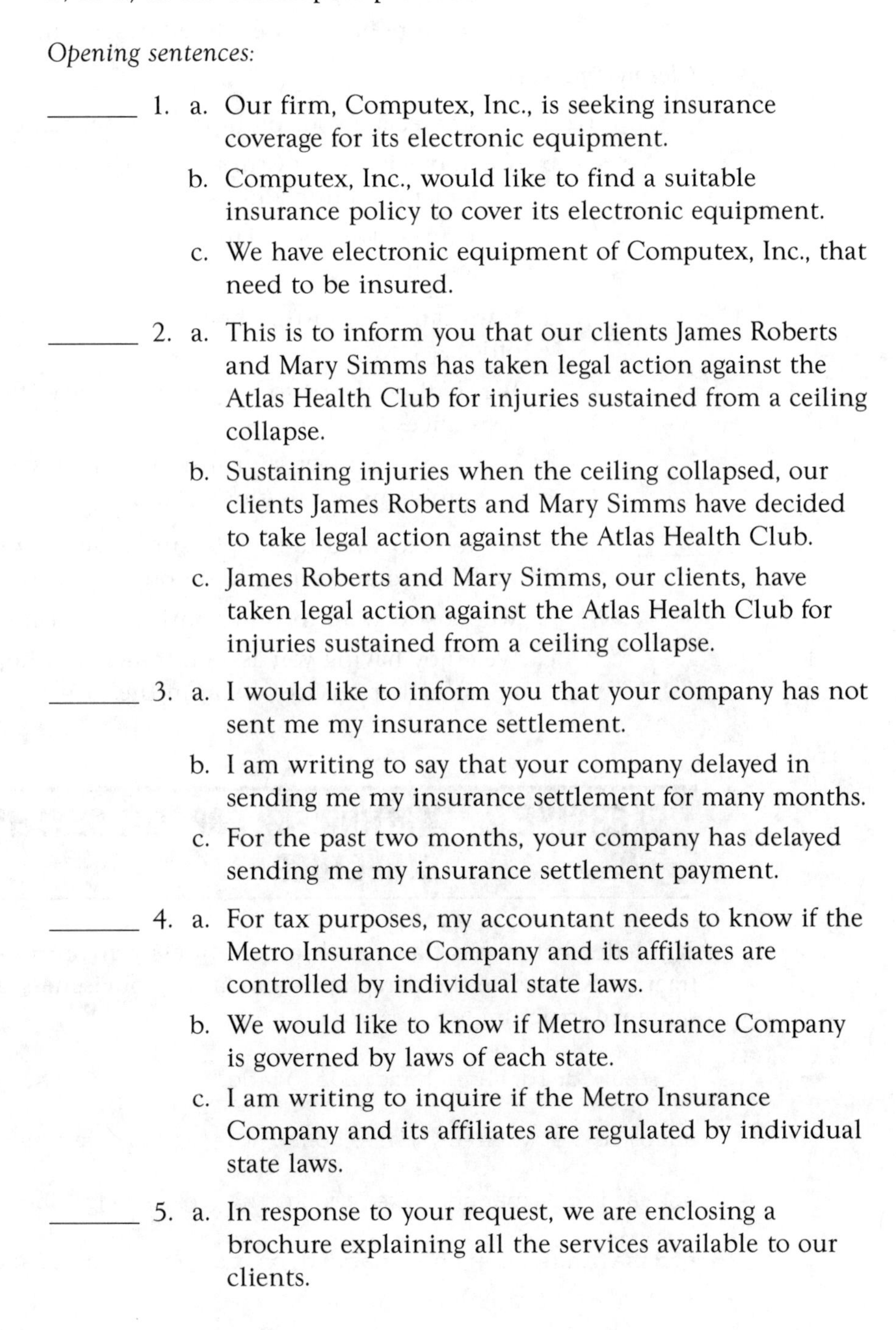

some sentences have built-in grammatical errors. Write your answer (A, B, or C) in the blank space provided.

Opening sentences:

_______ 1. a. Our firm, Computex, Inc., is seeking insurance coverage for its electronic equipment.

b. Computex, Inc., would like to find a suitable insurance policy to cover its electronic equipment.

c. We have electronic equipment of Computex, Inc., that need to be insured.

_______ 2. a. This is to inform you that our clients James Roberts and Mary Simms has taken legal action against the Atlas Health Club for injuries sustained from a ceiling collapse.

b. Sustaining injuries when the ceiling collapsed, our clients James Roberts and Mary Simms have decided to take legal action against the Atlas Health Club.

c. James Roberts and Mary Simms, our clients, have taken legal action against the Atlas Health Club for injuries sustained from a ceiling collapse.

_______ 3. a. I would like to inform you that your company has not sent me my insurance settlement.

b. I am writing to say that your company delayed in sending me my insurance settlement for many months.

c. For the past two months, your company has delayed sending me my insurance settlement payment.

_______ 4. a. For tax purposes, my accountant needs to know if the Metro Insurance Company and its affiliates are controlled by individual state laws.

b. We would like to know if Metro Insurance Company is governed by laws of each state.

c. I am writing to inquire if the Metro Insurance Company and its affiliates are regulated by individual state laws.

_______ 5. a. In response to your request, we are enclosing a brochure explaining all the services available to our clients.

b. We are pleased to send you the brochure you requested that describes our services.

c. A brochure has been enclosed describing our services.

Closing sentences:

_______ 6. a. We appreciate your prompt settlement of our claim.

b. We appreciate your firm's prompt action to make a settlement of our claim.

c. We express our appreciation of your prompt settlement.

_______ 7. a. If we can be of further help, feel free to contact our office.

b. We shall be happy to help you and be of further assistance.

c. Feel free to contact your insurance representative with any problems or questions you may have.

_______ 8. a. We look forward to the pleasure of serving you, and we hope you continue to do business with us.

b. We look forward to your continuing business.

c. We enjoy having you as a customer, and hope we can be of further service to you in the future.

EXERCISE 2 WRITING FOR CORRECT SENTENCE STRUCTURE

The letter below has flaws, such as wordiness, run-on sentences, and fragments. Rewrite the paragraphs for clarity, conciseness, and correct sentence structure.

You could save hundreds of dollars. By insuring your automobile with us this could mean savings for you. Our operating expenses are among the lowest in the industry and savings are then passed on to you, our customers.

We have additional money-saving discounts safe drivers, security systems, multi-car coverage. Along with our quality service which includes 24-hour access to our auto insurance specialists.

Call us free no obligation to get a rate quote.

Writing Pointers

Sales Letters

- Identify your potential customers.
- Capture reader's attention in the first paragraph.
- Use a question to get reader's attention.
- Specify details of a service or product.
- Give source for further information.
- Offer specific suggestions or recommendations for reader to act.
- Conclude on a positive note to promote goodwill.

Notification Letters to Customers

- State specific change (price increase, new location).
- State reason for change.
- Indicate benefit of change.
- Express sympathetic attitude toward hardships that might be created.

Model Sales Letter

Date

MS LAURA JACOBSON
11 THORNWOOD DRIVE
PARAMUS NJ 07652-3146

Dear Ms. Jacobson:

Question to arouse curiosity

Are you getting the best insurance rate for your safe driving record? Many companies insure drivers of all ages and levels of experience; they do not acknowledge good drivers. Costs of insurance claims are passed on to the insured.

Stresses benefits

Logical appeal to save money

Our insurance program is different. We, at Apex Insurance Company, primarily insure safe, experienced drivers with excellent driving records. As a result, we are able to keep our claim costs down and pass the savings to our customers. We have the lowest rates in the country. At the same time, our insurance package includes extra benefits not usually found in the industry.

Source for more information

Builds goodwill

Look at the enclosed brochure and learn more about our excellent coverage. Our experienced representatives will gladly answer any questions. We hope you will join our other subscribers with safe driving records and save hundreds of dollars a year.

Cordially yours,

Larry Ward
Vice President, Marketing

ri

Enclosure

Model Notification Letter

Date

MR EDWARD CARNEY
141 FT WASHINGTON AVENUE
ALBANY NY 12200-3541

Dear Mr. Carney:

SUBJECT: PREMIUM NOTIFICATION

Addresses specific issue

According to our records, you have not responded to our premium notification to continue your group life insurance policy. If payment of $73.50 is not received within ten days from the above date, we shall be forced to terminate your insurance. The insurance period covered under your renewal will be from July 1, 2005 to June 30, 2006.

Gives facts

Expresses sympathy

We are confident that this is an oversight on your part since this type of coverage is essential to your financial security and peace of mind. It is important to realize that the issue of money is not a concern during a crisis. You should be aware that you will be unable to re-enroll in this program if there is a lapse in the policy.

States benefit

Willingness to help

If you have any further questions or concerns, please feel free to contact our customer service department at 518-793-3000.

Sincerely yours,

Irene Dorsay
Customer Service Representative

ri

ACTIVITY F WRITING POWER

CASE STUDY 1: NOTIFICATION LETTER

Read Case Study 1. Then read Options A and B for each section of the letter. Select the option that you believe is most appropriate for the opening, body, and closing statements. Then key the entire letter based on the options you selected.

Case: A letter addressed to Mr. Richard Hodge from your insurance firm, General Casualty Corporation, tells him of an increase in his automobile premiums. Costs of auto repairs, legal fees, and medical care are on the rise; therefore, an increase of 5 percent is necessary. Explain that your company attempts to keep costs down. Ask the policyholder to cooperate by being a safe driver. The increase will take place when the insurance policy is renewed. Mr. Hodge's address is 22 Gates Street, New City, NY 10962–2245.

Opening Sentences:

a. General Casualty Corporation is committed to keeping your automobile insurance premiums low.

b. We at General Casualty Corporation are finding it hard to keep our costs low.

Option A or B _______

Body of letter giving specifics of situation:

a. We are faced with increasing costs from auto repair claims, legal fees, and medical expenses. Therefore, we have no choice but to impose a 5 percent increase on your premiums. This increase will take effect upon renewal of your policy.

b. When we reviewed our expenses, we found that claims from auto repairs, legal fees, and medical expenses are high and are going up even more. Therefore, we have no choice but to raise your premiums by 5 percent upon renewal of your policy.

Option A or B _______

Body of letter indicating company action, using positive tone:

a. We keep looking for new ways in our attempt to be efficient and to be able to keep our premiums at a reasonable cost to you. Please try and be a safe driver.

b. Our firm continuously studies ways to offer you efficient service at a reasonable cost. We ask that you assist by being a safe driver.

Option A or B _______

Closing statement to maintain goodwill:

a. Upon due date, you will receive your policy renewal.

b. For further information about your policy renewal, call us at (800) 750–3000. We value your continued patronage.

Option A or B _______

CASE STUDY 2: INFORMATIONAL LETTER

Read Case Study 2 and respond to the questions that follow the case study. Based on your responses, compose the letter to Mr. Steven Morris, 267 South Central Avenue, White Plains, NY 10602–9999.

Case: Your local office of the National Insurance Company is relocating to 2 Main Street, Tarrytown, New York 10591–4926. Draft a letter to your clients informing them of a change of address, which is a more convenient location and has parking facilities. You should also mention in your letter that a new service has been added at this new location. Insurance claims can be submitted here.

Organize Your Ideas

1. Indicate the most important point you want to make to your readers.

____

____

____

2. What should you include in the opening sentence?

____

____

____

3. What specific details should you include in the body of the letter?

____

____

____

4. What should you include in your closing statement that expresses a desire for a continued business relationship?

____

____

____

CASE STUDY 3: INFORMATIONAL FORM LETTER

Read Case Study 3, then develop an outline before composing the letter.

Case: Write a form letter from Jack Borden, Client Account Services, to the policyholders of Liberty Insurance Company telling them of the two new added services that are available: (1) round-the-clock service to file claims, renew policies, or ask for information (toll-free number 1–800–721–3000); (2) new web site to contact representatives (*http://www.Liberty.com*).

Appeal to clients that Liberty Insurance Company is committed to service. The address of Liberty Insurance Company is One Central Plaza, Washington, DC 20076–0023.

CASE STUDY 4: SALES LETTER

Read Case Study 4, then compose the letter.

Case: National Insurance Company is now offering a homeowner's policy to individuals who rent apartments. A form letter announcing this new type of coverage will be mailed to local residents. For the first year, the premium will be only $200 for all household contents up to $10,000 and will include $100,000 personal liability. If more coverage is needed, it can be obtained. Anyone interested in more information should complete the enclosed postage-paid card and return it to the office located at 2 Main Street, Tarrytown, New York 10591–4926. Use "Dear Neighbor:" for the salutation.

CASE STUDY 5: NOTIFICATION LETTER

Read Case Study 5; then compose the letter.

Case: Draft a letter to be sent to the policyholders who are renewing their automobile insurance. Inform them that there will be a 10 percent increase in their insurance rate. This increase is necessary because of the rise in claims on automobile accidents. Mention that one of the major causes of accidents is driving under the influence of alcohol. Remind the policyholders to be responsible drivers so that insurance rates will not continue to increase. Suggest the convenience of direct payment through the checking account. Conclude on a positive note by thanking them for choosing your firm, ABC Insurance Company. Your web site is *http://www.abcinsure.com*. For the salutation, use "Dear Policyholder." For the closing, use David Franke, Vice President of Operations.

ASSESSMENT: CHAPTER 2

English Power

Directions: *In each of the following sentences, underline the correct form of the verb so that it agrees with the subject.*

1. A new internship program (allow, allows) high school students to work with local insurance agencies to learn about the business.
2. Major changes (face, faces) the insurance industry during the next few years.
3. The enclosed package (provide, provides) you with a summary of all our insurance options.
4. With our insurance policy, the pedestrian (receive, receives) a fair settlement for injuries sustained in an accident.
5. Without adequate insurance, the flood victims (plan, plans) to apply for federal aid.
6. Management and union (have, has) agreed to offer group medical insurance at a discounted rate.
7. The $1 million umbrella clause (guarantee, guarantees) insurance payment for excessive claims.
8. In most insurance policies, you (is, are) required to meet the deductible sum before you (collect, collects) any reimbursement for your claim.
9. In many states, drunk drivers (lose, loses) their automobile insurance protection from an accident.
10. Major earthquakes and floods (deplete, depletes) an insurance company's financial resources.

Editing Power

Directions: *Edit, format, and proofread the following form letter for errors; key it correctly.*

Dear Policyholder:

National Insurance Company is opening a local office in your area. We are a nationwide insurance carrier with over $1 billion dollars in assets.

Our services includes insurance coverage for large cooporations as well as for individauls. You will find enclosed our booklet titled guide to Adequate insurance coverage. The booklet contain valuable information on types of insurance coverage. Our umbrela policy insure you for $2 million to cover possible catastrophic costs caused by a serious accident. To protect against auto theft, we encourage you to install anti-theft devices in your vehicle for which you will receive a 5 percent discount.

National also reward safe driving. You will receive a percent discount on your annual premum if you have a five-year safety record.

National insurance company is at you service. Please feel free to call us at (800) 424-3000.

Sincerely yours,

Mark Lewis

President
ri

Writing Power

Directions: *After reading the case below, draft a letter to Ms. Karen LaRubbia, 16 Villa Road, Santa Barbara, CA 93100–1050.*

A cover letter to Ms. LaRubbia will be attached to her automobile insurance renewal policy issued by National Insurance Company. The letter will express the company's appreciation for renewing the policy and will mention that a small increase in premiums was necessary because of accelerating claims. Clients should be informed that a 10 percent insurance discount will be given to drivers with a good safety record. Suggest in the letter that Ms. LaRubbia should review her coverage on bodily injury and property damage. The renewal form must be signed and returned within 15 days.

> "Coming together is a beginning, staying together is progress, and working together is success."
>
> *—Henry Ford*

3

Investing in Subject–Verb Agreement

WHAT YOU WILL LEARN IN THIS CHAPTER

- To apply correct subject–verb agreement principles in business correspondence
- To develop a business vocabulary in **FINANCE/INVESTMENTS**
- To proofread and edit correspondence and reports
- To compose an invitation letter
- To compose a letter of request
- To compose a letter of response to request
- To compose a promotional letter
- To write an e-mail message

Corporate Snapshot

MURIEL SIEBERT & CO., INC., bears the name of its founder. Although Muriel Siebert never graduated from college, she rose to partnership in a leading Wall Street brokerage firm. In a male-dominated industry, Ms. Siebert was willing to take risks and pursue her goals. In 1967, after many challenges, she was elected as the first woman member of the New York Stock Exchange. She spearheaded the drive to make her company a discount brokerage firm. Today Muriel Siebert & Co. is a successful and respected discount broker.

Acknowledging the financial needs of women, the firm began the Women's Financial Institute. This was the first organization within a brokerage house established to meet the growing demands of this segment of the population.

For Ms. Siebert, the secret of making a difference is to "take stands, take risks, take responsibility—and care deeply about how America's big institutions affect the lives of individual people." In keeping with this philosophy, the firm is involved in many philanthropic and educational programs.

Source: *http://www.msiebert.com*

ACTIVITY A LANGUAGE ARTS RECALL

Subject–Verb Agreement (continued from Chapter 2)

Rule 3.1: A phrase between the subject and verb does not affect agreement. The subject must agree with the verb.

Examples: The new ***booklet*** on financial opportunities ***is*** available.

Management, as well as the employees, ***is*** eligible for the stock-sharing plan.

Rule 3.2: If both of the subjects are singular and are joined by "either/or" or "neither/nor," the verb is singular.

Example: ***Either*** a savings account ***or*** a treasury bill ***is*** a good investment.

If both of the subjects are plural and are joined by "either/or" or "neither/nor," the verb is plural.

Example: ***Either*** savings accounts ***or*** treasury bills ***are*** good investments.

Rule 3.3: If one subject is singular and the other is plural, and the subjects are joined by "either/or" or "neither/nor," the verb agrees with the subject nearer the verb.

Example: ***Neither*** the stockholders ***nor*** the Executive Board ***agrees*** with their financing plan.

Are you sure ***either*** the controller ***or*** at least two members of the Board of Directors ***plan*** to attend the meeting?

Rule 3.4: Indefinite pronouns (each, every, anyone, everyone, someone, one, it) always require a singular verb.

Example: ***Everyone agrees*** that U.S. bonds are a safe investment.

Rule 3.5: When the word *there* begins the sentence, it is never the subject. The subject appears after the verb.

Example: There ***are*** many ***types*** of investment plans.

Rule 3.6: "The number" is a singular subject and takes a singular verb; "a number" is a plural subject and takes a plural verb.

Examples: ***The number*** of people investing in stocks ***is*** growing.

A number of banks ***have*** merged.

Check for Understanding

Directions: In the spaces that follow, indicate the subject with which the verb must agree and the verb in each of the following sentences. Also identify whether each subject is singular (S) or plural (P).

1. A glossary of financial terms appears in the back of the booklet. (3.1)
2. There are many advantages to an investment fund. (3.5)

3. According to reports, the number of investors in mutual funds is increasing every day. (3.6)
4. Either management or the employees are expected to attend the monthly meeting. (3.3)
5. Everyone on the research committee was asked to review the annual report. (3.4)
6. Each of us agrees that financial planning is important. (3.4)
7. Either savings bonds or a certificate of deposit is a wise choice. (3.3)
8. A growing number of employees have enrolled in company stock plans. (3.6)
9. Neither the stockbroker nor the financial advisor recommends a change in the client's investment portfolio. (3.2)
10. The number of firms offering employees stock option plans is increasing. (3.6)

	Subject	**Verb**	**S or P**
1.			
2.			
3.			
4.			
5.			
6.			
7.			
8.			
9.			
10.			

(Answers are given below.)

Answers to "Check for Understanding"

1. glossary, appears, S;
2. advantages, are, P;
3. (the) number, is, S;
4. employees, are expected, P;
5. Everyone, was asked, S;
6. Each, agrees, S;
7. certificate of deposit, is, S;
8. A (growing) number, have, P;
9. advisor, recommends, S;
10. (The) number, is increasing, S.

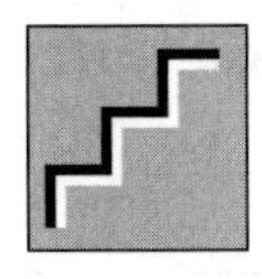

Building English Power: Subject-Verb Agreement

Directions: In each of the following sentences, underline the subject with a single line and underline the verb with a double line; in the space provided, change the subject and verb to plural form.

1. Foreign demand for American technology is increasing. (3.1)

   Plural Form: ______________________________

   ____

2. The number of stockholders is demanding more communication from corporate executives. (3.6)

   Plural Form: ______________________________

   ____

3. Neither the employee nor the manager agrees with the arbitrator's decision. (3.2)

   Plural Form: ______________________________

   ____

4. There is a concern about the company's financial condition. (3.5)

   Plural Form: ______________________________

   ____

5. The report in which we compare multinational corporations appears in several business magazines. (3.1)

   Plural Form: ______________________________

   ____

6. The enclosed pamphlet from the brokerage firm gives information on a variety of mutual funds. (3.1)

   Plural Form: ______________________________

   ____

7. Each of the brokers predicts an upswing in the stock market. (3.4)

   Plural Form: ______________________________

   ____

8. A seminar on financial planning is offered by the local bank. (3.1)

   Plural Form: ______________________________

   ____

9. There is a tax-free municipal bond that pays a high dividend. (3.5)

   Plural Form: ______________________________

   ____

10. Everyone agrees that individuals should have a savings plan. (3.4)

    Plural Form: ______________________________

    ____

11. Either a U.S. savings bond or a treasury note is a secure investment. (3.2)

    Plural Form: ______________________________

    ____

12. Each financial decision impacts your net worth. (3.4)

    Plural Form: ______________________________

    ____

13. The stock price of the electronics company has increased during the past year. (3.1)

    Plural Form: ______________________________

    ____

14. The fund, which invests in diversified stocks, has a high rate of performance. (3.1)

    Plural Form: ______________________________

    ____

15. Neither the bank nor the investment firm was willing to underwrite the project. (3.2)

    Plural Form: ______________________________

    ____

ACTIVITY B BUILDING WORD POWER

The words below are commonly used in the finance/investment industry. From this list, select the word or phrase that correctly completes each statement. Refer to the Personal Dictionary if you are uncertain of the meaning of a word.

blue chip stock(s)	prospectus
bonds	proxy
glossary	securities
investment portfolio	stockbroker
investments	stocks
multinational	web site
mutual fund	

_______________ 1. Companies often borrow money by issuing _______, which are loans to be repaid on a certain date, including interest.

_______________ 2. _______ is another term used for stocks and bonds.

_______________ 3. _______ are shares that individuals own as an investment in companies. These shares are sold by companies to raise capital.

_______________ 4. An investor usually refers to a(n) _______ to examine the detailed financial background of a proposed investment in a firm.

_______________ 5. Individuals attempt to make more money through _______ in securities, savings, and real estate.

_______________ 6. An investor has a(n) _______ that lists the securities owned by the individual.

_______________ 7. An individual investor may buy shares in diversified securities offered by an investment group. This type of investment is known as a _______.

_______________ 8. ______ people who want to buy or sell securities contact a(n) _______ who is a registered sales representative of a member firm in the stock exchange.

_______________ 9. To understand the terms used in the financial pages of a newspaper, refer to a(n) _______ for an explanation of the words.

_______________ 10. Corporations that do business in foreign countries are known as _______ firms.

_______________ 11. The shareholder was unable to attend the annual meeting and voted by _______ for the proposal on performance-based compensation.

_____________ 12. For the latest stock information, visit the _______ *http://www.dailystocks.com*.

_____________ 13. IBM, General Electric, and Exxon Mobil are examples of _______.

On the Alert

- a/an (indefinite article)—"An" is used before all words beginning with a vowel. It is also used before the silent "h." "A" is used before all other words.

  Examples: ***a*** number; ***an*** individual; ***an*** hour

- advice (noun)—recommendation

  Example: The financial advisor gave her clients ***advice*** on municipal bonds.

- advise (verb)—to recommend or give counsel

  Example: The broker will ***advise*** his investors to attend the seminar on the new stock offering.

- personal (adjective)—individual, private

  Example: The bank approved a ***personal*** loan to purchase the car.

- personnel (noun)—staff, employees

  Example: All ***personnel*** in the accounting department are being trained to use the new software program.

ACTIVITY C INDUSTRY PERSPECTIVES: INVESTMENTS

Directions: Using appropriate revision marks, proofread and key the following report.

Rule Reference	
2.1	A successful investor need to be well informed about
3.5	financial issues. There is many ways to obtain relevant
	information and advise about investing in securities.
3.1	The financial sections of a newspaper offers its
	readers valuable data on current economic conditions and
2.1	investment opportunities. Investment columns often

appears in newspapers and magazines written by experts
3.6 who offer money management suggestions. A number of
television programs focuses on stock market news,
securities investments, and industry information.
2.1 Internet systems offers subscribers up-to-the-minute
stock informationa, and daily trading activities of
2.2 securities and mutual funds is available instantaneously.

In a marketplace enviromnent, known as the stock
2.2 exchange, the buyng and selling of securities occurs.
Only registered brokers can trade on the stock exchange.
A individual who is interested in investing in stocks
and bonds should have a knowledgeable broker.

The most popular and largest exchange is known as the
1.4, 1.7 New york stock Exchange. Every leading american
3.4 industrial, financial, and service corporation trade its
1.5 stock on the NYse.

3.6 The number of individuals investing in securities are
3.2 growing. Neither a stockbroker nor a financial advisor
are solely responsible for an individual's investment
decisions. Choices must be based on personnel financial
2.2 goals and information that has been researched.

Did you find these errors?
10 subject-verb agreement
3 typographical/spelling
4 capitalization
3 confusing words
1 grammar

ACTIVITY D EDITING POWER

Task A: Edit, format, and proofread the following letter. Use proofreaders' symbols to mark the errors, then key it correctly.

Errors
- Subject–verb agreement
- Capitalization
- Typographical
- Confusing word

MS BARBARA SIMMONS

ONE PARK ROAD

KANSAS CITY MO 64141-6202

Dear Ms. Simons:

A growing number of americans is using financial planning to manage there money better. Everyone agree that investors should be well informed of market trends. We invite you to send for a free copy of our Financial planning Guide. it is written in an easy, understandable manner. Ten experts in the investment field offers suggestions on how to invest money. A glossary of financial terms appear in the back of the booklet for reference. There is many types of investments. We suggest that you carefully read what the recognized authorities in the financail field have to say. Then you will be able to make wise decisions.

Why not send for your copy today?

Sincerely yours,

Suzanne Rudman

Vice President
ri

Task B: Edit, format, and proofread the following unarranged letter for all errors and missing notations, then key it correctly.

MR TROY JOHNSON, 22 CARA LANE, CINCINNATI OH 45230-7214

Dear Mr. Johnson:

We have enclosed a copy of our current prospectus which serve as your reference for information about the World Fund.

The primary objective of our money fund is to invest in foreign companies that have a potential for capital growth. This fund focuses on the emerging asian market. Everyone agree that this is a promising area for business expansion.

World fund is recognized for its quality service. Our financial consultants are always ready to provide you with the personnel attention and advise you deserve. Neither institutions nor a small investor are excluded from participating in the fund to receive more information or to open a account, call us at 1-800-567-9000 or visit our web site http://www.worldFund.com.

Sincerely yours, Fred Rodriguez, Account Executive

ACTIVITY E WRITING PRACTICUM

Good writing involves the flowing of ideas so that the thought in the first sentence leads to the very next idea in the sentence that follows. In essence, ideas are interconnected within a reading passage.

EXERCISE 1 REARRANGING SENTENCES FOR CONTINUITY OF IDEAS

In problems 1 through 5, rearrange the sentences in a logical order, list the sentence order in the spaces provided, then key each passage.

1. a. The use of computers to keep abreast of financial information is becoming increasingly popular.

   b. Online computer networks are making conventional ways of researching information as obsolete as using a typewriter.

   c. With the Internet, an investor can communicate with investment specialists, retrieve securities information, and refer to financial databases.

   d. All you need are a personal computer, a modem, and Internet access to keep the information superhighway at your fingertips.

   List in order of preference ____ ____ ____ ____

2. a. Last summer, I had experience working as an intern for a small investment firm in my hometown.

   b. In response to your advertisement placed in the summer job mart at River Community College, I would like to apply for the position.

   c. I was fascinated by the challenges of the brokerage field.

   d. Please send me an application for your management trainee program.

   List in order of preference ____ ____ ____ ____

3. a. There are no fees for this service, and you can terminate the plan at any time.

b. If you have any further questions, please call your account representative.

c. With this plan, you authorize us to transfer a fixed amount each month from your bank account to your mutual fund account.

d. You will be pleased to know that we are introducing an automatic monthly investment plan.

List in order of preference ____ ____ ____ ____

4. a. We have enclosed a brochure that explains our services; feel free to contact one of our experienced representatives for further help.

   b. All of our investment consultants are professionals who are knowledgeable in the securities market.

   c. Our firm offers a wide range of investment plans tailored to the needs of our clients.

   d. Thank you for contacting Investment Services Corporation.

List in order of preference ____ ____ ____ ____

5. a. As an experienced investor, you will find this monthly newsletter an incredible source of information on all types of investments.

   b. We have enclosed your free copy of *Financial Review.*

   c. Each month a prominent business professional writes about a financial topic of interest to our readers.

   d. Please take time to read this valuable pamphlet, and don't hesitate to send in your subscription.

List in order of preference ____ ____ ____ ____

EXERCISE 2 INAPPROPRIATE PHRASES

At times, we use slang or inappropriate language in our daily speech. However, when more formal communication is used, such as in the writing process, some phrases are unacceptable. The chart on the following page lists common examples.

Unacceptable Phrases	Acceptable Phrases
being that	because
due to the fact that	due to/because
got to	must
hadn't ought to	should not
kind of	somewhat
lots of	many
on account of	since
per	according to
should of	should have
at this point in time	now

Directions: Substitute the correct word(s) for each of the underlined inappropriate phrases.

1. There are lots of good investment opportunities for individuals who do careful research of the financial market.
2. Per your telephone conversation, the investors' meeting will be held next month.
3. At this point in time, you are able to purchase stock on the Internet.
4. Due to the fact that you have an excellent credit rating, we are increasing your cash limit to $10,000.
5. The manager should of discussed the stock option plan with the staff beforehand.
6. Being that you have been with the firm for one year, you are eligible to join its investment plan.
7. On account of the company's high earnings for the year, the stock has increased in value.
8. The president of the company hadn't ought to announce his successor until the board of directors meets next week.
9. When you go for the interview, you got to know the company's history.
10. We were kind of disappointed in the seminar on investments.

Writing Pointers

Letter of Invitation

- Mention purpose of invitation.
- Include specific information—time, place, location, and topic.
- Express benefit to reader.
- Indicate where further information can be obtained.
- Request confirmation or response.

Letter of Request

- Be specific in your request—date, place, item.
- Get to the point; ask for specific information.
- Give reason for request.
- Address the letter to an individual whenever possible.
- Show appreciation.

Letter of Response to a Request

- Acknowledge appreciation for interest in product or service.
- Give information pertinent to the request; don't give needless information.
- Indicate action to be taken.
- Offer further assistance to the reader.

Letter Format

- Subject line is double spaced after the salutation. (See model.)

Model Invitation Letter

Date

MS KAREN CHAN
22 EAST STERLING DRIVE
BALTIMORE MD 21201-6347

Dear Ms. Chan:

SUBJECT: ANNUAL MEETING OF SHAREHOLDERS

Purpose of invitation / *Specific information* / *Response requested*

Apex Electronics Company will hold its Annual Meeting of Shareholders on Wednesday, May 10, 2006, at the Convention Center in Orlando, Florida. We look forward to your attending in person; otherwise, complete and return the enclosed proxy within two weeks. If you plan to attend, check the box on your proxy card so that we may send you an admission invitation.

At this year's meeting, the program includes the following agenda:

Agenda topic specified

1. Election of Board of Directors
2. Approval of appointment of independent auditors
3. Ratification of stockholders' proposals that are discussed in attached statement

Further assistance offered

If you have any further questions regarding the proposals to be presented at the meeting, feel free to call us at 1-800-289-4000.

Sincerely yours,

Sylvia Torres

ri

Enclosure

Model Letter of Request

Date

MR RICHARD BARNES
PORTFOLIO MANAGER
ASSET MANAGEMENT FUNDS
1170 PIKES PEAK ROAD
BOULDER CO 80323-5400

Dear Mr. Barnes:

Positive opening

Praises reader

Your speech at the Investors' Convention in Salt Lake City was most informative. I found the booklet, *Cybermarket for Investors,* which you distributed at the conference, an invaluable reference guide.

Reason for request

Specific request

This guide would be an excellent resource for my staff to share with their clients. Is it possible to receive 50 copies? If there is a charge, please let me know.

Positive appeal to reader

Since this reference guide is an extremely valuable book for investment specialists, I hope you will be able to fulfill this request.

Sincerely yours,

Lisa Willis
Vice President

ri

Model Response to a Letter of Request

Date

MS LISA WILLIS
PARKER FINANCIAL GROUP
55 THORNWOOD ROAD
DES MOINES IA 50392-4150

Dear Ms. Willis:

Acknowledge appreciation

Positive comments about convention

Thank you for your thoughtful comments regarding my recent presentation in Salt Lake City. I, too, found the convention both enjoyable and informative. All the sessions were most professional.

Indicates action

Yes, we will honor your request and ship the 50 copies of *Cybermarket for Investors* next week. It has been a very successful booklet, and I am sure your staff will find that it is an excellent guide when dealing with clients.

Gives pertinent information about positive response

Offers assistance

Although the booklet is free, we will bill you for the shipping charges. If there are any further questions, please feel free to contact me.

Sincerely yours,

Richard Barnes
Portfolio Manager

ri

ACTIVITY F WRITING POWER

CASE STUDY 1: PROMOTIONAL LETTER

Read Case Study 1. Then read Options A and B for each section of the letter. Select the option that you believe is most appropriate for the opening, body, and closing statements. Then key the entire letter based on the options you selected. Address the letter to investors.

Case: General Investors, an investment firm, is introducing an automatic monthly investment plan for mutual fund accounts. With this plan, an investor who has a savings or checking account will be able to ask the bank to transfer to General Investors a specific amount of money each month. The plan will remain in effect until the investor decides to cancel. It is an easy and systematic way to save. Let the investor know that an account representative is available to answer all the investor's questions.

Opening sentence

a. General Investors is delighted to let you know that it is starting an automatic monthly investment plan for mutual fund customers.

b. Are you interested in an automatic monthly investment plan for your mutual fund account?

Option A or B ______

Sentences in body of letter

Sentence 1:

a. Under this plan, you inform your bank to transfer money each month from your savings or checking account to this account.

b. The plan now available with General Investors authorizes your bank to transfer each month from your savings or checking account a fixed amount of money to your account with us.

Option A or B ______

Sentence 2:

a. This plan remains in effect until you decide to cancel. It is an easy, systematic way to increase your mutual fund.

b. This plan will continue until you tell us to stop withholding money. Now you have an easy way to save money.

Option A or B ______

Closing sentence

a. If you have any questions, please call your account representative.

b. Also, if you need more information, call us and we will be happy to answer your questions.

Option A or B ______

CASE STUDY 2: RESPONSE TO A REQUEST

After reading the case, respond to the questions that follow to help organize your ideas before beginning to compose the letter. Then draft the letter to:

MR MICHAEL GIBSON

85 BYRON PLACE

NEW HAVEN CT 06515–2281

Case: Mr. Gibson has called your office requesting information on one of your mutual funds, the Select Fund. Prepare a cover letter to send with the literature and prospectus that describe this fund. You may want to mention to Mr. Gibson the advantages of this fund, which has had an average growth rate of 5 percent annually. Also, there are no sales commissions. State that there is a toll-free number for further information about our services, (800) 531-5575.

Organize Your Ideas

1. Indicate the most important point you want to make to Mr. Gibson.

____

____

2. What should you include in the opening sentence?

____

____

3. What specific details should you include in the body?

____

____

____

4. What should you include in your closing statement that would get a response from your reader?

____

____

CASE STUDY 3: INVITATION FORM LETTER

Read Case Study 3, then develop an outline before composing the letter.

Prepare a form letter to investors inviting them to attend a seminar on "Investing for the 2000s." The speaker, Melissa Hodge, is a well-known financial planner. She will offer guidelines for preparing an investment portfolio. Free time will be available for questions and answers. The seminar will be held on Thursday, June 30, 20___, at 8 p.m. at the Savoy Plaza Hotel. Invite interested investors to return the reply card immediately.

CASE STUDY 4: REQUEST LETTER

Read Case Study 4; then compose the letter.

The investment firm of Brooks and Carney plans to expand its financial services. To determine which services are important, the firm has designed a survey questionnaire to be sent to its clients.

You have been assigned the task of composing a letter for your boss, Daniel Brooks, president, in which you ask clients to complete the survey.

To get a good response, you will have to explain the importance of the survey and the need for clients' participation. Include a subject line referring to the Important Client Survey.

Address the letter to Mr. Alan Barnett, 82 Cimarron Drive, Oklahoma City, OK 73100–2290.

CASE STUDY 5: INVITATION LETTER

Read Case Study 5; then compose the letter.

Case: You are drafting a form letter to be sent to prospective clients. The letter invites them to subscribe to The Price Fund. It is a no-load fund that invests in blue chip stocks and bonds. For the past five years, the fund has outperformed other funds in its class. Last year it had a 20 percent return. Mention that the success of the fund is due to the company's philosophy of disciplined investment in growth stocks. Experienced financial specialists are always available to help clients with financial questions. A prospectus is available. They can visit the web site: http://www.PriceFund.com or contact a representative at 1-800-620-2000.

Reminders: Use an opening that will attract the reader's interest. For the salutation, use "Dear Prospective Client." The letter should be signed by Maria Salina, Chief Marketing Officer.

E-Mail Correspondence

One of the most widely used features of the Internet is electronic mail. It has become a cost-effective means of communication within an organization as well as with customers. E-mail facilitates the flow of information among correspondents at various locations and time zones. Individuals receive messages more quickly than the printed document and are able to respond immediately.

An advantage of electronic mail is that you can communicate with an individual halfway around the world. The person may be from a different culture, have different customs, and speak a different language. You need to be sure that your message is understood so that no misinterpretations occur.

Although e-mails may be informal among friends, in business correspondence e-mails must be correct, clear, concise, and easy to understand. They should basically follow the same rules of good writing as in traditionally written correspondence.

Model E-mail

	Subject of E-mail
	File Edit View Tools Message Help
	Reply Reply All Forward Print Delete Previous Next
Receiver's e-mail address	To: cruz@aol.com Cc:
Purpose of message	Subject: Subscription Renewal Attachments: none
Salutation included; message sent outside company	Dear Ms. Cruz: We acknowledge your subscription renewal to Money Matters. Every Friday online you will receive a weekly update of financial news.
Message is concise and clear	Thank you for renewing.
Contact information	Laura Crawford Subscription Manager lcrawford@aol.com

E-Mail Format (may vary depending on the software)

To: The receiver's e-mail address (some individuals prefer to add their names to the e-mail address; place it in angle brackets)

From: The sender's e-mail address (may be omitted)

Cc: Copy sent to other receivers

Subject: Heading should give reader immediate information related to the content

Closing: may be optional; however, in some e-mails it is unclear who the sender is. Therefore, individuals may prefer to include their name, title, company e-mail address, and phone number

Body of e-mail: For easier readability use bullets, listings, and double space between paragraphs

Basic rules in e-mail correspondence:

- Subject line is important in an e-mail; it should inform the reader of the content.
- Message should be focused, concise, and clear.

- Message needs to be error-free (correct grammar, punctuation, and spelling).
- Message should be proofread.
- Correspondence should not be long (recommended maximum 250 words).
- E-mails are not confidential; do not send security-sensitive information without protective measures.
- The use of attachments is not encouraged—people are reluctant to open attachments; however, there are times when it is necessary to attach a report.
- The script of an e-mail should not be in all capital letters.
- Slang and sarcasm should be avoided.

CASE STUDY 6: E-MAIL

Read Case Study 6; then complete the e-mail.

Send an e-mail to Ms. Kim Lang (Klang@worldnet.com) inviting her to attend a seminar on mutual fund investing. The seminar will be held on June 10 at the offices of American Financial Corporation, One Wall Street, NY.

Inform Ms. Lang that this seminar will help her make sound financial decisions. If she wishes to attend, Ms. Lang should make a reservation. Your firm's web site is *http://www.afc.com.*

This e-mail is being sent by Wendy Palez, financial advisor. The subject is Seminar on Mutual Fund Investing.

E-Mail Format

In each chapter, use the following format when writing the heading of an e-mail.

To: (Receiver's e-mail)
Cc: (optional) (Copies sent to other receivers.)
Subject: (Heading related to e-mail content)

ASSESSMENT: CHAPTER 3

English Power

Directions: *In each of the following sentences, underline the correct form of the verb so that it agrees with the subject.*

1. There (is, are) databases that review the past performance of mutual funds.
2. The Securities and Exchange Commission, a governmental agency, (monitor, monitors) the stock brokerage firms.
3. The board of directors of the company (recommend, recommends) a change in the employee stock purchase plan.
4. Either our management team or our stock specialists (examine, examines) each company's growth potential.
5. The number of people joining stock investment clubs (is, are) increasing.
6. The annual meeting of shareholders (provide, provides) for an exchange of views between management and investors.
7. The mutual fund, which invests in diversified stocks, (rank, ranks) high in long-term performance.
8. A number of stockbrokers (advise, advises) their clients to review carefully their financial goals before purchasing securities.
9. Municipal bonds, which are often tax-free, (offer, offers) investors an alternative means of saving.
10. Each of our analysts (has, have) been assigned a specific industry to follow.

Editing Power

Directions: *Edit, format, and proofread the following form letter for errors; key it correctly.*

Dear Investor:

Universal, Inc., are committed to provide its customers with the best possible service at the lowest cost. Our experienced representatives, who are specialists in the securities market, assumes responsibility for certain clients.

The number of people investing in stocks are growing. Our diversified types of accounts, such as mutual funds, municipal bonds, and tax-deferred plans, meets the personnel needs of all investors. Our state-of-the-art communications and computer technology serves our domestic and foreign markets.

The enclosed prospectus present details of our complete range of savings and investment opportunities. We suggest that you read this material carefully. Each of our representatives stand ready to advice you.

Thank you for your interest in Universal, Inc.

Sincerely yours,

Henry Walls

Assistant Vice President

Writing Power

Directions: *After reading the case below, draft a form letter to clients of your stock brokerage firm.*

This letter should announce to your firm's clients that a new service will begin on December 1 to provide state-of-the-art communications. A toll-free telephone number, (800) 223–5495, will be available 24 hours a day, seven days a week, so that clients may speak to an account executive. Clients may contact a representative for securities information to purchase stocks and bonds or to receive updates on their account balances. Assure the clients that your firm will continue to provide the best possible service.

> "Leadership has a harder job to do than just choose sides. It must bring sides together.
>
> *Jesse Jackson*

Telecommunicating with Verbs

WHAT YOU WILL LEARN IN THIS CHAPTER

- To use correct verb tense in business correspondence
- To develop a business vocabulary in **TELECOMMUNICATIONS**
- To proofread and edit correspondence and reports
- To rewrite a poorly written letter of apology
- To compose a memorandum
- To compose a letter of complaint
- To write an e-mail memorandum

Corporate Snapshot

INTEL ® Corporation is the world's largest chip maker, supplying advanced technology solutions for the computing, networking, and communications industries. Intel's chips can be found in everything from PCs to servers, cell phones, and telecommunications equipment.

Through Intel Involved, the company's worldwide volunteer program, employees focus their energies on being an asset to the communities where they live and work. The company has a strong commitment to education. The Intel Innovation in Education initiative contributes more than $100 million annually in more than 30 countries around the world to strengthen mathematics and science education, the foundation for the high technology industry and for much of the global economy.

Source: Intel Corporation, Corporate Affairs, Media Relations

ACTIVITY A LANGUAGE ARTS RECALL

Verb Tense

Verb tense refers to time. The verb in a sentence indicates when the action occurred.

Rule 4.1: simple present—actions that are happening now.

Rule 4.2: simple past—actions that happened in the past and were completed.

Rule 4.3: simple future—actions that will happen.

Simple Present Tense	Simple Past Tense	Simple Future Tense
(verb)	(verb+"d" or "ed")	("will" or "shall"+ verb)
invite	invited	will invite

Perfect tense indicates an activity that happened in the past, either before the present or before another time in the past, or an activity that will have been completed before a time in the future. The perfect tense uses the auxiliary (helping) verb "to have" plus a special form of the main verb, known as the past participle, usually formed by adding "d" or "ed" to the regular verb. The three perfect tenses are present perfect, past perfect, and future perfect.

Rule 4.4: present perfect—action began in the past but was completed either before the present or at some unspecified time in the past; when used with "since" or "for," the action is continuing.

Rule 4.5: past perfect—action began in the past and was completed before another time in the past.

Rule 4.6: future perfect—action will be completed before another time in the future.

Present Perfect Tense	Past Perfect Tense	Future Perfect Tense
active voice: subject performs the action		
(have, has)	(had)	(will have, shall have)
have (has) invited	had invited	will have invited
have (has) chosen	had chosen	will have chosen
("Chose" is an irregular verb.)		
passive voice: subject is acted upon		
has been invited	had been invited	will have been invited

Check for Understanding

Directions: Compare each set of the verbs in Column A and Column B; then answer the questions below.

Column A	Column B
invite	has invited
discuss	have discussed
produce	will have produced
analyze	had analyzed
speak	had spoken

Question: What do all of the verb forms in Column B have in common?

____

Conclusion: Complete the following sentence.

To form the perfect tense of regular verbs, ______________________

____

(Answers are given below.)

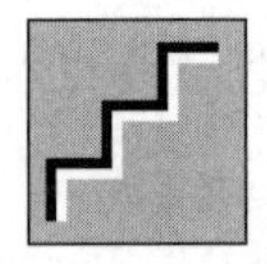

Building English Power: Verb Tense

Directions: In the spaces provided on page 83, indicate the subject and verb in each of the following sentences. Also identify the tense of each verb.

1. Last year, our office introduced electronic mail for speedier communications. (4.2)
2. Fax machines have decreased in price over the past year. (4.4)
3. During the past decade, consumer interest in cellular phones has increased. (4.4)
4. The corporation will have introduced videoconferencing by next year. (4.6)
5. The editor of the magazine has used desktop publishing for the past year. (4.4)
6. By the year 2025, senior citizens will have outnumbered teenagers by two to one. (4.6)
7. The control of computer viruses has been discussed at our staff meetings. (4.4)
8. With a CD-ROM, you have the opportunity to listen to music and study the composition. (4.1)
9. A task force of advertisers has been organized to study how high-tech media can reshape ads. (4.4)
10. A new manager will be hired by next month. (4.3)
11. The company had planned to upgrade its computer capabilities to include high-speed data networking. (4.5)
12. Many executives carry a notebook computer when traveling in business. (4.1)

Answers to "Check for Understanding"

Consist of auxiliary verb (has, had) plus main verb with "d" or "ed."

To form the perfect tense of regular verbs, use the auxiliary verb with the past participle of the main verb.

13. At last month's meeting, the personnel directors had discussed the difficulties in hiring qualified technicians. (4.5)
14. We had demonstrated our graphics software at the trade show before our competitor announced a similar product. (4.5)
15. The bank has encouraged its customers to use the Internet for onscreen checking transactions. (4.4)

	Subject	Verb	Verb Tense
1.			
2.			
3.			
4.			
5.			
6.			
7.			
8.			
9.			
10.			
11.			
12.			
13.			
14.			
15.			

ACTIVITY B BUILDING WORD POWER

The words below are commonly used in the telecommunications industry. From this list, select the word or phrase that correctly completes each statement. Refer to the Personal Dictionary if you are uncertain of the meaning of a word.

configuration
electronic mail (e-mail)
facsimile equipment (FAX)
Internet
laptop computer(s)
modem
monitor
mouse
networking
notebook
protocols
software
telecommunications
videoconferencing

_______________ 1. A popular method of holding face-to-face meetings between locations by using video equipment is known as ________.

_______________ 2. With the need to send printed copy instantaneously, the ________ is replacing conventional mail deliveries.

_______________ 3. To receive up-to-date stock information, you can add a(n) ________ to connect a PC to a telephone to transmit information.

_______________ 4. When traveling, business executives carry portable computers that are called ________.

_______________ 5. ________ is the term used to describe the transfer of data from one place to another over communication lines.

_______________ 6. ________ allows businesses to send, receive, and store messages over telecommunications facilities.

_______________ 7. A more compact, lightweight computer than a laptop is known as a(n) ________.

_______________ 8. The computer ________, which is the display screen, may produce a glare that is disturbing to the user.

_______________ 9. Many computer users prefer to use a(n) ________ to move the cursor around the screen freely without using a keyboard.

_______________ 10. The engineer is studying the ________ of the office before submitting a new design plan.

_______________ 11. A new ________ program allows you to do your banking by computer.

_______________ 12. Through the ________, an individual can communicate electronically worldwide on many subjects and draw information from databases.

_______________ 13. To communicate successfully on the Internet, standardized procedures and regulations, known as ________, need to be used.

_______________ 14. ________ systems can connect computers within an office and throughout the world.

On the Alert

- its (possessive pronoun)—belonging to

  Example: The computer in my office is useless because ***its*** monitor is broken.

it's (contraction)—it is

Example: ***It's*** a high priority for a business to subscribe to an Internet service.

- past (noun)—the time before the present

Example: In the ***past***, electronic games did not exist.

past (adjective)—no longer current

Example: The firm is studying its ***past*** sales records in the Northeast.

passed (past tense of verb)—moved on; accomplished

Example: Most of the students ***passed*** the course.

- affect (verb)—to influence

Example: Advances in telecommunications ***affect*** people's daily lives.

effect (noun)—result

Example: One of the ***effects*** of the computer has been a need to retrain employees.

effect (verb)—to bring about

Example: The law that was passed in Congress ***effects*** a change in taxes.

- assistance (noun)—help, aid

Example: ***Assistance*** was provided to the fire victims.

assistant (noun)—an individual who assists someone in a managerial position

Example: The office ***assistant*** will provide you with the necessary data.

- site (noun)—location

Example: The ***site*** of the new building is a popular tourist attraction.

sight (noun)—1. the ability to see

Example: He lost his ***sight*** after the accident.

2. (noun) something worth seeing

Example: The lake in the winter was a beautiful ***sight***.

sight (verb)—to see or observe

Example: Columbus ***sighted*** land in 1492.

cite (verb)—to quote or mention

Example: When you do research, ***cite*** the source of quoted material.

- percent—Express percentages in figures; write out the word "percent"

  Example: 5 percent

ACTIVITY C INDUSTRY PERSPECTIVES: TELECOMMUNICATIONS

Directions: Using appropriate revision marks, proofread and key the following report.

2.1 The fast-paced telecommunications industry have affected every aspect of our lives both at home and in business. What is telecommunications? Its the process of transmitting information over long distances through a communication line such as telephone, cable, microwaves,
1.3 and satellites. this information may take the form of voice, data, image, or message.

3.1, 2.1 Expedient transmission of information and data on a global scale are crucial in today's business markets. Whether a firm's offices are located in the United
1.4 States, europe, or asia, business transactions can take place anywhere and at any time. With telecommunications, business is less dependent on face-to-face interactions. Business dealings can be accomplished through networking
2.1 systems. Within minutes, a sender connect with the

receiver to share information and data. This new
2.1 technology also enable employees to interact with there
coworkers anywhere in the world. Therefore, it is
3.4 essential that everyone respect different cultures and
customs of individuals in foreign countries.

On a personnel level, individuals have become more
dependent on telecommunications. The cell phone today is
viewed as a necessity to keep in touch with family and
3.6, 2.1 friends. The number of people surfing the web for the
latest information on consumer goods, medical data, and
3.4 financial news are constantly growing. Everyone agree
that in this Digital Era technological advances will
continue to effect the way we live and do business.

Did you find these errors?
4 subject-verb agreement
4 word usage
3 capitalization

ACTIVITY D EDITING POWER

Task A: Edit, format, and proofread the following letter for verb tense, subject–verb agreement, and other errors. Use proofreaders' symbols to mark the errors, then key it correctly.

Errors
- subject-verb agreement
- verb tense
- confusing words
- typographical

MR. EDWARD BURKE, JR.

1100 BROAD STREET

CLEVELAND, OH 44190-8030

Dear Mr. Burke:

Thank you for your inquiry about our computer product line. The authorized dealer nearest you is Computer Mart, 5 Metro Office Park, Cleveland, OH 44190-3598. Computer Marte offer a complete line of products from laptop computers to network systems. For businesses, custom configurations and on-sight installations are available. You may want to consider Computer Mart's excellent leasing program, which has been popular for for many years.

Our dealer has maintain it's efficient and reliable service for the passed 20 years. The sales representatives has received training on our equipment. Therefore, they will be able to discuss your needs and recomend equipment best suited for your business.

Feel free to visit the showroom; you will find everyone most helpful.

Sincerely yours,

Lloyd Pearson
Sales Manager

Task B: Edit, format, and proofread the following memorandum for all errors, especially verb tense, then key it correctly. (See Appendix B for memorandum format.)

This memorandum is being sent from Inez Santos, Manager of the Marketing Division, to the Sales Staff. The subject is New Product.

```
(P) Our latest upgraded computer, Bell SX25, is
available on December 1. As we have discuss at previous
sales meetings, this product will have consumer appeal.
It has been equipped with a CD-ROM storage device and
multimedia software. It also has been preload with
several computer programs. Enclosed you will find a
booklet describing the the new product. (P) The next
sales meeting held in the Main Building Conference Room
on october 10 at 9:15 a.m. The sales approach to
introduce this new product will have been discussed at
this meeting. (P) If you are unable to attend, please
call my office to arrange for a video conferencing
session.
```

ACTIVITY E WRITING PRACTICUM

EXERCISE 1 CHANGING VERB TENSE FROM PRESENT TO PAST

The following fax was sent to every branch office of a large corporation. The message was written in the present tense. Rewrite the fax and change all of the verbs to the simple past tense. Use correct format.

DATE: (CURRENT DATE)

TO: BRANCH OFFICES

R. J. TOMPKINS CORPORATION

FROM: RAY DEJESUS

SUBJECT: TELENETWORK SYSTEM

We want to alert all of our branch offices that we are converting our telephone lines to a new computerized system on September 5.

This allows for several improvements over our old system. Each new desk unit consists of a modular console that includes fax, voice mail, and three-way conferencing features. Executive offices have hookups for videoconferencing and taping facilities. We are confident that our interoffice communications benefit from this change.

The plan is to install the system over the Labor Day weekend so that the business day has few interruptions.

We appreciate your cooperation in this matter.

EXERCISE 2 CHANGING VERB TENSE FROM PRESENT TO PRESENT PERFECT

In the following paragraphs that will be published in a magazine column, change the verb tense from simple present to present perfect tense.

The Internet introduces a new era in communications. Through a computer network, friends, business colleagues, and customers interact and share information instantly with one another. The Internet is as popular as using a cellular phone or faxing a message. It is the focus of the computing world.

Computer networks change the way business handles its transactions. Information is globalized and resources are available from all over the world. The following services are available to the Internet user: online shopping, banking transactions, checking status of orders, researching databases, voice messaging, and exchanging e-mail.

The Internet is growing rapidly. Protocols for this online computer network do not limit the type of users. An increasing number of large corporations, small businesses, and individual subscribers join the Internet for their means of communication.

Writing Pointers

Letter of Complaint by Customer

- Describe specifically the item or model.
- State the problem clearly and reasonably.
- Avoid using offensive or unnecessary words.
- Indicate your suggestion for resolving the problem.
- Avoid threats and demands.

Letter of Apology

- Explain reason for delay or error.
- State action taken.
- Express empathy.
- Offer assistance or an alternative solution.
- Avoid negative words that suggest blame (regret, unfortunately, fault).
- End with goodwill statement.

Memorandum

- Use for internal communications.
- Make it brief and specific.
- State purpose of memorandum in opening paragraph.
- Provide a clear, informative subject line.
- See Appendix B for memordum format.

Model Letter of Complaint

Date

MR ROY FOLSOM
PRESIDENT
CENTRAL TECHNOLOGY COMPANY
881 CENTRAL AVENUE
AKRON OH 44321-1452

Dear Mr. Folsom:

Problem stated clearly

One month ago, I ordered ten mobile phones from your firm, which still have not arrived. Your salesperson promised delivery within two weeks. When I spoke to your customer representative about this problem, I was told that the delay was a backlog of orders on this item.

Polite approach Expresses concern regarding delivery

This excuse is unsatisfactory. I can understand a delay; however, no attempt has been made to explain when the order will be filled. These mobile phones are essential to my business; without them, I cannot contact my employees to service my customers.

Requested action

Please let me know the status of this order and how you intend to resolve this matter.

Sincerely yours,

Kenneth Murdoch
Manager

ri

Model Letter of Apology

Date

MR KENNETH MURDOCK
MANAGER
THE ELECTRONIC STORE
224 CLEARVIEW PARKWAY
RICHMOND VA 23234-5512

Dear Mr. Murdock:

Apology for delay

Opening shows empathy

Explanation given

We are sorry that your shipment of ten mobile phones has been delayed. An unexpected equipment failure has caused a backlog of this item. Since this particular model is not readily available from other manufacturers, we are unable to fill the order. Our representative should have explained the situation to you.

States action to be taken

We understand your frustration and will make every attempt to special handle the order when we receive the mobile phones. A shipment is expected next week.

Positive tone

Gesture of goodwill

Please accept our apologies. A 10 percent deduction will be credited to your account; we hope this meets with your approval. You may contact me at any time.

Sincerely yours,

Roy Folsom
President

ri

ACTIVITY F WRITING POWER

In Chapters 4 through 7, Case Study 1 will consist of a poorly written letter. After reading the case study, critique the letter by responding to the questions. This will guide you in writing an effective letter by adding, rewording, or deleting words or phrases that do not demonstrate good writing techniques. Remember to use the six Cs of writing a good letter: **clarity, correctness, conciseness, completeness, consistency,** and **courteousness**. Review the "Writing Pointers" in each chapter to help you improve your writing skills.

CASE STUDY 1: REWRITE A POORLY WRITTEN LETTER OF APOLOGY

After reading the poorly written letter addressed to one of your customers, Cyber Mart, Inc., 11 Elwood Drive, Portland, ME 04101-3190, your supervisor, James Winter, requested that you rewrite the letter. To help you complete this task, (a) answer the questions that follow; (b) rephrase sentences, if needed; and (c) rewrite the letter for mailing.

Case:

Dear Mr. Ruiz:

Our company is sorry to say that we have had problems in our company and cannot ship at the present time the five laser-jet printers you ordered.

Unfortunately, the reason for this delay was due to a job slowdown at the manufacturing plant. We recognize that the delay will be an inconvenience to you. However, this situation was beyond our control.

Once again, we are sorry. We hope to serve you in the future.

Organize Your Ideas

1. In this letter of apology, what is the first statement you should make to your customer?

 ____

 ____

2. What is wrong with the second sentence?

 ____

 ____

3. How can you improve the third and fourth sentences to reflect a caring attitude and to show empathy for the customer?

 ____

 ____

4. What is missing in this letter that should be included?

 ____

5. What is wrong with the closing sentence?

 ____

 ____

CASE STUDY 2: MEMORANDUM ON PRICE INCREASE

After reading the case below, respond to the questions that follow, which will help you organize your ideas. Then prepare a memorandum to the sales force from Glen Chase, marketing vice president, on the price increase. Use correct memorandum format.

Case: Glen Chase, marketing vice president, has been asked to notify his sales force that a 5 percent increase in price on the entire line of electronic equipment will become effective on June 1. The price increase is the result of rising manufacturing costs. A revised price list has been attached to this memo. Inform the salespeople that they should contact their managers if they have any questions.

Organize Your Ideas

1. What is the most important point you want to make to your sales force?

 ____

 ____

2. What would you include in the opening sentence?

 ____

 ____

 ____

3. What specific details would you include in the body of the memorandum?

 ____

 ____

4. What would you say in your closing statement that would get a positive reaction from your sales force?

 ____

 ____

 ____

CASE STUDY 3: LETTER OF COMPLAINT

Read Case Study 3; then develop an outline before composing the letter.

Case: You purchased a cellular telephone five months ago from CompuMax, an authorized dealer of Cellular Systems. The model you bought, XL95, has a one-year warranty from Cellular Systems, the manufacturer. Recently, you began to hear a high-frequency hum when using the phone. This is distracting and interferes with your conversation. Notify the manufacturer that you are returning the phone for a replacement. The manufacturer's address is 17 Grand Avenue, Des Moines, IA 50312–9526.

CASE STUDY 4: LETTER OF APOLOGY

Read Case Study 4; then complete the letter.

As a customer representative, you have the responsibility of answering complaint letters. Ms. Esther Reynolds of 21 Magnolia Boulevard, Savannah, Georgia 31504–5213, is complaining that her recent telephone bill shows charges for long-distance calls that she did not make. She feels that there must be some mistake and asked you to check her records. You reviewed this matter and found Ms. Reynolds is correct. Write a letter explaining the situation and apologizing for the error.

CASE STUDY 5: LETTER OF COMPLAINT

Read Case Study 5; then complete the letter.

Sid Kantor, manager of operations, has asked that you draft a letter to Frank Ambrosio, vice president of sales, on a matter of the unsatisfactory operation of new printers Tech 2005. Five printers were purchased last month; three of them are not producing quality prints: After 100 copies, the colors fade so that the photo is no longer bright or sharp. In your business, these prints are unacceptable. The service person who inspected the three printers concluded that there was nothing mechanically wrong. You depend on good equipment to get satisfactory results. This is a serious matter for your firm. Request that the printers be replaced. You expect a prompt solution to the problem. Address the letter to Mr. Frank Ambrosio, Vice President of Sales, Tech, Inc., 3211 Office Park, Cincinnati, OH 45200.

E-Mail Memorandums

E-mails are replacing the use of paper memos within a company. E-mail memos offer several advantages:

- Correspondence can be sent quickly for an immediate reply.
- Memorandums frequently are written in an informal manner suitable for e-mail.
- Correspondence can be sent at any time. The mail is held in an electronic mailbox until the recipient opens it.
- Messages can be sent to several people in the organization at one time.
- Reports and other documents from other programs can be attached.

Model E-mail Memorandum

Subject of Email

File Edit View Tools Message Help

Reply Reply All Forward Print Delete Previous Next

To: O'Brien.sue@packards.com
Cc:
Subject: Meeting for Sales Staff
Attachments: none

Informal style

Salutation and closing optional

Just a reminder that Wednesday's meeting to discuss the sales promotional campaign for the Notethink laptop is scheduled for 10 a.m. See you then.

CASE STUDY 6: E-MAIL MEMORANDUM

Read Case Study 6; then complete the e-mail.

Send an e-mail memo to the Human Resources manager, Marta Santiago (martainteractive@opcit.com). Inform her that you would like to meet with her on June 10 at 9 a.m. to discuss management's plan for implementing a company policy for the personal use of the Internet and e-mail. Explain to her that the company is aware that this is a growing problem among the employees who see no wrongdoing, especially when the computer is used at the end of the workday. Indicate to Marta Santiago that you would like to hear suggestions and recommendations that relate to this issue and that employees would accept as a new policy.

Suggestion: You may want to include reasons why management is concerned with the problem.

The subject of this e-mail is Meeting to Discuss Internet and E-Mail Usage.

ASSESSMENT: CHAPTER 4

English Power

Directions: *In each of the following sentences, change each italicized verb to the tense indicated in parentheses.*

1. Everyone *agree* that technology *affect* job skills. (simple past)

   ____

2. The recent poll *indicate* that the number of individuals who work at home *is increasing*. (present perfect) ______________

   ____

3. Advanced technology *allow* management to conduct global videoconferencing. (present perfect) ______________

   ____

4. Directories of companies *be* available on the Internet. (simple future) ______________

   ____

5. The researcher *compile* the information quickly from the database of cases. (past perfect) ______________

   ____

6. By the end of the year, the firm's branches *receive* the in-house company newsletter. (simple future) ______________

   ____

7. Our new buyers' directory *serve* as an excellent reference guide to technology products. (past perfect) ______________

   ____

8. The CD-ROM drive *increase* the capabilities of the personal computer. (present perfect) ______________

   ____

9. The vendor *introduce* a new software package to erase computer viruses. (simple past) ______________

   ____

10. During takeoff and landing, airlines *prohibit* the use of cellular phones and laptop computers. (present perfect) ______________

____

Editing Power

Directions: *Edit, format, and proofread the following form letter for errors, then key it correctly.*

Dear Customer:

As a valued customer of our products, you are invite to take part in a new support program at no cost to you.

In the passed, many of you have express frustration when problems has arisen with your computer equipment. A hotline (800-758-0023) has been install to route you directly to our staff, who can answer your toughest questions. If you have difficulty and need personnel assistants, one of our local specialists will be sent immediately to your sight to survey the problem. If you want to expand your computer systems, our specialists will arrange to give a demonstration to your staff on some of our other products, such as videoconfrencing systems, notebook computers, and flat-panel desktop monitors.

```
You have been a favored customer of ours, and we thank

you for your continued support.

Sincerely yours,

Gil Wager
Vice President
ri
```

Writing Power

Directions: *After reading the case below, draft a memorandum to sales representatives from Thomas Saunders, Manager. The subject is "Staff Meeting."*

Thomas Saunders wants to notify the sales representatives that a meeting will be held next Monday (insert correct date) at 9 a.m. in the conference room. The agenda of the meeting will be the price increase for the entire line of computers due to escalating manufacturing costs. This price increase will take effect at the beginning of next month. The staff should be prepared to offer suggestions on ways to maintain goodwill when informing the customers about the price increase.

> "Alone we can do so little; together we can do so much."
>
> *Helen Keller*

5

Lots of Commas in Real Estate

WHAT YOU WILL LEARN IN THIS CHAPTER

- To identify compound sentences
- To punctuate compound sentences with a comma
- To develop a business vocabulary in **REAL ESTATE**
- To proofread and edit correspondence and reports
- To compose collection letters at various stages
- To compose a request letter
- To compose an informational letter
- To write an e-mail memorandum

Corporate Snapshot

Malls: An American Phenomenon

The automobile has had a profound impact on the American lifestyle. In the 1920s, shopping centers with plazas surrounded by stores with parking spaces began to appear.

When Americans began to leave the cities to settle in suburban areas, retail stores also began moving from downtown areas. The regional mall was introduced in the 1950s. Shoppers began to travel distances to malls that were anchored by department stores, specialty shops, and a central pedestrian walk condutive to browsing. The shift of retail businesses from city centers to suburban areas had taken place.

Gradually developers expanded regional malls into the popular mega malls. The most recent real estate concept is the upscale mall with a cityscape environment that includes walkways, flowers, fountains, as well as restaurants, office space, and apartments.

Are malls headed for extinction? Probably not as long as architects and real estate developers can create space that contributes to a good quality of life.

ACTIVITY A LANGUAGE ARTS RECALL

Sentence Structure: Compound Sentence

A **compound** sentence has two or more independent clauses that are connected by a coordinating **conjunction** (and, or, but, nor).
As you learned in Chapter 1, an independent clause consists of a subject and a verb with its modifiers. The independent clause must make sense grammatically.

Comma–Conjunction

Rule 5.1: Use a comma before a conjunction that joins two independent clauses.

Example: The application must be completed by the borrower, ***and*** it will be processed as soon as it is received.

Rule 5.2: When a subject does not follow the conjunction, no comma is needed.

Example: The application blank must be completed by the borrower ***and returned*** within ten days. (The subject is missing after the conjunction "and.")

Rule 5.3: When the conjunction "and" is followed by the word *that,* do not place a comma before the conjunction unless the word *that* is the subject of the clause.

Example: The real estate broker believed the property was priced at market value ***and that*** a sale would occur shortly. The mortgage was approved by the bank, ***and that*** expedited the closing date for the house. (The word *that* is the subject.)

Check for Understanding

Directions: Compare the sets of sentences and answer the questions below.

1. a. We will be happy to meet with you and discuss possible sites.

   b. We will be happy to meet with you, and our salesperson will discuss possible sites.

2. a. Our representatives have expertise in the field of real estate and charge reasonable fees.

   b. Our representatives have expertise in the field of real estate, and they charge reasonable fees.

3. a. Our real estate office has relocated to a site that is much larger and more convenient for our customers.

   b. Our real estate office has relocated to the Downtown Mall, and that is a site our customers find more convenient.

Questions

1. What do sentences 1b, 2b, and 3b have in common?

____

2. What is the difference between sentences 1a and 1b?

   ____

3. Which other two sentences have this same difference?

   ____

4. Why isn't there a comma before the word *that* in sentence 3a?

   ____

(Answers are given below.)

Conclusion: Complete the following sentence.

When a ________ sentence contains two or more _______ clauses, there is a _______ before the conjunction.

Answers to "Check for Understanding"

1. Each sentence has two independent clauses. 2. Sentence 1a has only one independent clause; sentence 1b has two independent clauses. 3. Sentences 2a and 2b. 4. The word *that* is not the subject of the second clause. (Rule 5.3)

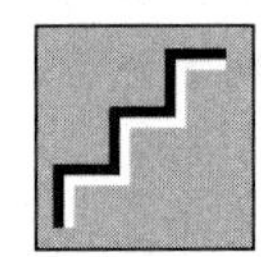

Building English Power: Comma-Conjunction

Directions: Insert commas where required. If no comma is needed, write "C" (correct) in the space provided.

______________ 1. The home improvement business has become a major industry and future trends indicate a growing market. (5.1)

______________ 2. The custom-built homes are designed by a well-known architect but the prices are moderate. (5.1)

______________ 3. The residential community is located near the highway and is convenient to major shopping malls. (5.2)

______________ 4. The five acres of land will be sold to individual buyers or to a real estate developer. (5.2)

______________ 5. Mortgage rates have been low during the last year and that will affect future sales. (5.3)

______________ 6. Office space rents by the square foot and is quite costly in desirable business districts. (5.2)

____________ 7. Waterfront property has become increasingly expensive but buyers are always interested in purchasing such real estate. (5.1)

____________ 8. The tenant owes three months' rent and that is the reason the landlord will take legal action. (5.3)

____________ 9. Congratulations on your election to the governance committee of the building and the board thanks you for your contributions to good management. (5.1)

____________ 10. The architect's design for the neighborhood park has been completed and construction should begin in two months. (5.1)

____________ 11. The land will be sold by the acre and parceled out in smaller lots. (5.2)

____________ 12. The property has easy access to the downtown area and that site is located right off Exit 23 on the interstate highway. (5.1)

____________ 13. The court ruling prohibits the use of wetlands for commercial development but the firm will appeal the case. (5.1)

____________ 14. A day-care center will be built in the new corporate park and it will be available to employees and the community. (5.1)

____________ 15. The Association of Realtors will hold its annual conference in June and we urge individuals who are interested in the field of real estate to attend the meeting. (5.1)

ACTIVITY B BUILDING WORD POWER

The words below are commonly used in the real estate industry. From this list, select the word or phrase that correctly completes each statement. Refer to the Personal Dictionary if you are uncertain of the meaning of a word.

condominium	periodically
cooperative	principal
corporate park	prorating
debris	Realtors
delinquent	references
index	security deposit
lease	town code
mortgage	

_______________ 1. Before approving a loan, a bank secures ________ on the borrower's credit rating and character.

_______________ 2. Home builders must follow government regulations, which are written in the ________.

_______________ 3. ________ that accumulates during construction must be discarded according to environmental regulations.

_______________ 4. Fire inspectors ________ inspect buildings for fire-code violations.

_______________ 5. Interest rates on adjustable-rate loans are based on a(n) ________, which is a mathematical calculation that determines the amount a borrower pays each month.

_______________ 6. The ________ on a mortgage is the total amount borrowed to purchase property.

_______________ 7. A tenant is usually required to sign a legal agreement with the landlord, known as a(n) ________, which states the terms and conditions of rental.

_______________ 8. A(n) ________ is a stipulated amount of money given to the landlord by the tenant to cover potential damages or nonpayment of rent.

_______________ 9. The architect plans to design a(n) ________, which will be a cluster of office buildings in a park-like setting.

_______________ 10. Individuals known as ________ are experts in the field of buying and selling real estate.

_______________ 11. Most individuals will need to obtain a(n) ________ from a bank or other lending institution to finance the cost of purchasing a house.

_______________ 12. A(n) ________ offers an individual the opportunity to own a unit within a multi-dwelling complex. A monthly maintenance fee covers the cost of common facilities.

_______________ 13. A form of property ownership in some cities is the purchase of shares in a corporation that owns the building and the land. The shareholder owns the apartment. This type of real estate is known as a(n) ________.

_______________ 14. The method used to divide utility and heating bills equally over a 12-month period is known as ________.

_______________ 15. Customers who are ________ in paying bills are sent monthly reminders to settle their charges.

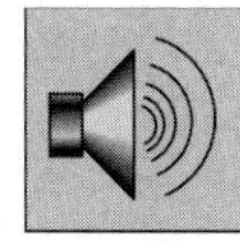

On the Alert

- of (preposition)—used to indicate relationship

  Example: He is president ***of*** the Real Estate Board.

- off (adverb)—not on

  Example: Shingles were blown ***off*** the roof.

- perspective (noun)—a point of view

  Example: The guide gave an account of the lifestyles of the early 1900s from an historical ***perspective***.

- prospective (adjective)—pertaining to the future

  Example: The ***prospective*** buyer filed an application for a mortgage.

- principal (noun)—

  1. the head of a school
  2. a sum of money owed, not including interest

  Example: The school ***principal*** repaid the ***principal*** on her loan in 24 months.

- principle (noun)—a basic truth

  Example: The landlord's renting practices were based on unfair ***principles***.

ACTIVITY C INDUSTRY PERSPECTIVES: HOME MORTGAGES

Directions: Using appropriate revision marks, proofread and key the following report.

Rule Reference 5.3

```
    Purchasing a home is one of the most important
financail decisions you will make. Shopping for a home
mortgage can be confusing and that often frustrates a
perspective buyer. The real cost of a house is the
purchase price plus the financing charges. These charges
```

2.1, 5.1 often amounts to twice the cost of the house and the buyer should shop around for the best morgage deal.

There are key words that a borrower should understand when applying for a mortgage. The two most common key
2.1 words is fixed rate and adjustable rate. With a fixed-rate mortgage, the monthly payments of principle and
3.1 interest never changes; and the debt is paid of at the end of the term.

The interest rate on an adjustable mortgage can
5.2 change periodically, and is based on an index or formula. A borrower must be aware that the interest rate can increase or decrease during the life of this type of mortgage.

Frequently, a buyer is charged points on the mortgage. This means that the borrower pays the lender a set fee that is based on the amount of the mortgage.

Buyers should compare different mortgage plans that meet there financial needs.

Did you find these errors?
3 punctuation
3 subject–verb agreement
4 word usage
3 typographical/spelling

ACTIVITY D EDITING POWER

Task A: Edit, format, and proofread the following form letter. Be aware that commas have been inserted incorrectly in several places. Use proofreaders' symbols to mark the errors, then key the letter correctly. Use "**Dear Neighbor:**" as the salutation for this letter.

Errors
- commas
- subject–verb agreement
- word usage
- typographical

Our company, Capital Realtors, have been engaged in the Westchester real estate industry for over 25 years and our firm have earned a fine reputation in the community. We take pride in analyzing a perspective client's needs, and matching the proper house to the client. Our experienced brokers carefully research each property but the cost is never past on to the buyer. We believe that the market in this area is recovering, and that there are many good opportunities to buy real estate. If you are interested in either a new home or a parcel of land for development, visit our office and speak to one of our experienced representatives. Satisfied clients are the best source of references and a list of repeat clients are available to you.

We look forward to serving you. Don't hesitate to call Capitol Realtors at (914) 623-1000.

Sincerely yours,

John Costello

Vice President
ri

Task B: Edit, format, and proofread the following unarranged letter for all errors. Supply missing parts and paragraph indentations; key it correctly. (See Appendix B for a model of a listing.)

This letter is being sent from James Hoffer, Managing Agent, to Mr. and Mrs. David Ewing, One Circle Drive, Boston, MA 02106–3540.

We have examine your plans for the renovation of your kitchen and they have been approved by us. The following procedures must be observe: 1. Contractors must submit certificates of insurance before work begins. 2. All electrical and plumbing work must be performed by licensed persons and must comply with town codes. 3. You are liable for any damages to electrical and telephone wiring that may occur during construction and all costs will be billed to you. 4. Removal of old appliances and debris are you responsibility but we can arrange for pickup. If you have any questions, please call me at my office.

ACTIVITY E WRITING PRACTICUM

The following exercise will check your knowledge of correct use of adjectives and adverbs. Review the rules before you complete the exercise.

Adjectives and Adverbs

Rule 5.4

- **Adjectives** modify or describe nouns or pronouns. Adjectives may add clarity or additional information to a sentence.

Example: The ***large*** house is built in a ***wooded*** area of the ***suburban*** town. (descriptive)

- Adjectives are used to compare items. To most adjectives, add "er" when comparing two items; add "est" to compare three or more items.

Examples:	**Comparative**	**Superlative**
large	larger	largest
warm	warmer	warmest
tall	taller	tallest

- To compare adjectives of three or more syllables or words ending in "able," use "more" or "most."

Example: The new site is a ***more desirable*** location than the original proposal. (comparative)

The building will be constructed in the ***most desirable*** part of town. (superlative)

- Adjectives "a," "an," or "the" are also referred to as *articles*. "A" and "an" are general; "the" is specific.

Use **a** when the word that follows begins with a *consonant* or the sound of a long *u*.

Use **an** when the word that follows begins with a *vowel* or silent *h*.

Examples:		
	a skyscraper (consonant)	an attorney (vowel)
	a house (consonant)	an hour (silent *h*)
	a university (long *u*)	an uncle (vowel)

- When regular verb forms are used as adjectives, add "d" or "ed."

Examples: The ***enclosed*** check is for payment of last month's rent.

- **Adverbs** modify verbs, adjectives, or other adverbs. They tell when, where, how, and how much. Frequently, adverbs are formed by adding "ly" to an adjective.

Examples: This ***beautifully*** decorated home is featured in a national magazine. (*Adverb* "beautifully" *modifies adjective* "decorated.")

The management team ***unanimously*** agreed to landscape the property.

(*Adverb* "unanimously" *modifies verb* "agreed.")

The remodeling of the home progressed ***very*** slowly.
(*Adverb* "very" *modifies adverb* "slowly.")

EXERCISE 1 USE OF ADJECTIVES AND ADVERBS

Complete each sentence with the correct form of the word in parentheses. Write the adjective or adverb form in the space provided.

_____________ 1. The ________ date for a site inspection has been postponed. (schedule)

_____________ 2. The ________ report on real estate investments is available in my office. (complete)

_____________ 3. The ________ parties met to reach an agreement. (interest)

_____________ 4. The community envisions a small park in the ________ populated area. (dense)

_____________ 5. You performed ________ on the real estate licensing examination. (satisfactory)

_____________ 6. The ________ signatures were missing from the ________ contract. (require, revise)

_____________ 7. The ________ land could be used only for commercial purposes. (lease)

_____________ 8. The ________ renovated building will be open this weekend for prospective tenants. (new)

_____________ 9. The developer anticipated major renovations to the ________ ________ properties. (recent, acquire)

_____________ 10. The ________ broker will get you a ________ good purchase price on the vacant building. (experience, fair)

EXERCISE 2 COMPARATIVE AND SUPERLATIVE FORMS; USE OF "A" "AN"

Choose the comparative or superlative form (as indicated in parentheses at the end of each sentence) of the word in parentheses or correct use of "a" "an" write the answer in the space provided.

_____________ 1. The (large) one-bedroom apartment will become vacant next month. (superlative)

_____________ 2. A (more, most) favorable city in which to relocate our offices is Seattle. (comparative)

_____________ 4. One of the (more, most) popular tourist attractions is Disneyland. (superlative)

_____________ 5. The property manager gave a (more, most) informative presentation than the other speakers. (comparative)

_____________ 6. The (more, most) elegant furnishings in the estate will be auctioned. (superlative)

_____________ 7. The (more, most) successful managers have people skills. (superlative).

EXERCISE 3 CORRECT WORD CHOICE

Choose the correct word in parentheses and write the answer in the space provided.

_____________ 1. (Most, Almost) of the reports show an increase in the sale of private homes.

_____________ 2. (Most, Almost) half of the respondents favor the construction of a town park on the vacant lot.

_____________ 3. We received (bad, badly) news about the proposed variance to build a garage.

_____________ 4. He spoke (bad, badly) about the proposed building code.

_____________ 5. The students performed (good, well) at the state competition.

______________ 6. (A, An) added bonus will be free rent for one month.

______________ 7. (A, An) unique feature of the house is the shape of the swimming pool.

______________ 8. The judge gave (a, an) unfavorable ruling on the housing project.

______________ 9. The gardener is paid at (a, an) hourly rate.

Active/Passive Voice

- Use the **active voice** of the verb to strengthen a message. In an active sentence, the doer of an action, the subject, precedes the verb. The *subject* of the sentence *performs* the action.

  Example: The bank approved the home mortgage for the young couple.

- The **passive voice** indicates who or what is receiving the action. A sentence in the passive voice is generally considered weak because the *subject receives* the action rather than performing the action. It is occasionally used for emphasis, tact, and sensitivity.

  Example: The young couple's home mortgage was approved by the bank.

EXERCISE 4 CONVERTING FROM PASSIVE TO ACTIVE VOICE

Strengthen the following sentences by changing them from passive voice to active voice.

Example: Your rent for the month of September ***was paid*** by John. (passive voice)

John ***paid*** your rent for the month of September. (active voice; subject is doer of action)

1. Fifty single-family homes will be built by the local construction firm.

____

____

2. An increase in real estate taxes has been proposed by the city council.

____

____

3. A reduction in the price of the new homes was announced by the builder.

____

____

4. The plan for urban development was submitted to the town board by the engineer.

____

____

5. Mortgage financing is offered to first-time home buyers by Union Bank.

____

____

6. A plan to convert the building to cooperative ownership has been submitted by the landlord.

____

____

7. Renovations were vetoed by the Board of Directors.

____

____

8. The mortgage application that was submitted by the young couple was rejected by the bank.

____

____

9. The difficulty in changing the existing building code was mentioned by the speaker.

____

____

10. The annual meeting of the condominium was attended by the majority of the shareholders.

____

____

Collection Letters

To collect overdue accounts, businesses use collection letters. They are written in several stages from sensitive reminders to a stern tone that indicates readiness to take action. The first stage is a simple reminder to pay the bill. The letter is written in a tone that maintains goodwill and at the same time obtains a positive response.

At the second stage, the customer still neglected to pay the bill. The letter is written in stronger language persuading the customer to make payment. At this stage, the letter may indicate a willingness to help the customer meet the financial obligation.

The final stage is where action needs to be taken. The letter is firm and explains the consequences if the bill remains unpaid.

Writing Pointers—Collection Letters

First Stage—Reminder

1. Remind customer of overdue payment in a cordial tone.
2. Appeal to customer's responsibility for payment.
3. Assume person intends to make payment but has forgotten.

Second Stage—Request

1. Remind customer of failure to respond to first reminder.
2. Ask if a problem exists in paying the bill.
3. Suggest a willingness to cooperate in a plan for making payment.
4. Communicate your desire to receive payment promptly or to hear from the customer.
5. Refer to the consequences of nonpayment (late charge, poor credit rating).

Final Stage—Action

1. Convey stern tone in demanding payment.
2. Stress consequences of nonpayment, which can result in poor credit rating.
3. Threaten customer with turning the matter over to a collection agency or taking legal action.
4. Send letter certified, return receipt requested.

Letter Format

For letters addressed to companies, the salutation is either "Gentlemen" or "Ladies and Gentlemen."

Model First Stage of a Collection Letter

Date

MR SAM BURNS
11 BOULEVARD AVENUE
BOSTON MA 02175-3342

Cordial tone — Dear Mr. Burns:

Gentle reminder — This letter is a reminder that your rent for the month of August is past due. We are sure that this is an oversight. Your prompt payment of $725 would be appreciated.

Goodwill — Keep up your good credit rating.

Sincerely yours,

Rose O'Malley
Building Manager

ri

Model Second Stage of a Collection Letter

Date

MR SAM BURNS
11 BOULEVARD AVENUE
BOSTON MA 02175-3342

Dear Mr. Burns:

Reminder

Perhaps you have overlooked our first reminder that the rent for your apartment is still unpaid. As of this date, you are two months behind in rent, totaling $1,450. As a valued tenant, you must have a reason for this nonpayment.

Appeal to self-respect

Refer to consequences of nonpayment

To avoid an unfavorable credit rating, please call our office at 617-567-9832 within the next few days to discuss the matter. If you are unable to make full payment, we shall try to work out a satisfactory solution.

Communicate willingness to cooperate

Offer solution

Your cooperation will be appreciated.

Sincerely yours,

Rose O'Malley

ri

Model Final Stage of a Collection Letter

Date

CERTIFIED MAIL

MR SAM BURNS
11 BOULEVARD AVENUE
BOSTON MA 02175-3342

Dear Mr. Burns:

Severe tone

Stress consequences

Your rent is now five months overdue. We sent you several reminders but no response has been received. The delinquent stage of your account leaves us no alternative but to begin eviction proceedings.

Give time limits

We must hear from you or receive a check for $3,625 within the next ten days; otherwise, your account will be forwarded to our attorney for legal action.

Maintain goodwill

We are confident that you will resolve this matter.

No negative language

Sincerely yours,

Rose O'Malley
Building Manager

ri

ACTIVITY F WRITING POWER

CASE STUDY 1: REWRITE OF POORLY WRITTEN COLLECTION LETTER, SECOND STAGE

Rewrite the following poorly written collection letter. It is a second stage reminder letter to Ms. Lauren Foley, One Wood Avenue, New City, NY 10956–1008 about her past-due mortgage payment.

To help you complete this task, (a) answer the questions that follow, (b) rephrase sentences, if needed, and (c) rekey the letter for mailing.

Case

Dear Ms. Foley:

Mortgage payment for the months of September and October are late. We sent you a previous letter reminding you of the situation. There is a penalty of $50 for this delayed payment. Please be so kind to send us a check for $1,150 to cover the two months. If we do not hear from you, we will be forced to take legal action.

Questions

1. What is the overall tone of this letter?

____

2. What should the writer's main goal be for a second reminder letter?

____

3. What is wrong with the last sentence?

____

____

4. What is a closing sentence that would get results?

____

____

CASE STUDY 2: COLLECTION LETTER, FINAL STAGE

After reading this case, respond to the questions that follow, which will help you organize your ideas. Then draft a letter.

Case: You are the managing agent of a building at 150 Poe Lane, St. Louis, MO 63100–4509, in which Mr. and Mrs. Richard Greene reside. After a reminder sent to the Greenes that their rent is three months past due, the rent remains unpaid. In your letter, request that they pay their total bill of $2,700. Payment must be received immediately or management will be forced to take legal action.

Organize Your Ideas

1. At what stage of the collection process should you write this letter?

____

2. What is the most important point that you want to make to Mr. and Mrs. Greene?

____

____

3. What would you include in the opening sentence?

____

____

4. What specific details should you include in the body of the letter?

____

____

____

5. What would you include in the closing statement to get a positive reaction from the tenant?

____

____

____

CASE STUDY 3: COLLECTION LETTER

Read Case Study 3; then develop an outline before composing the letter.

Case: Ms. Eileen Robb, a shareowner at Rockledge Cooperative Apartments at 11 Rockledge Road, Hartford, CT 06180–1254, refuses to pay a $25 late penalty fee for nonpayment of her monthly maintenance charge. According to the rules of the cooperative, any shareowner who does not pay the monthly charges on the first of each month will be charged a penalty.

Ms. Carole Ponte, managing agent, has been asked to send a letter to Ms. Robb requesting that this fee be paid immediately; otherwise, the amount will be doubled after ten days.

CASE STUDY 4: LETTER OF REQUEST

Read Case Study 4; then compose a letter.

Case: Mr. and Mrs. John Tisch, who reside at 2 Maiden Lane, Tarrytown, NY 10532–0050, are planning to expand their kitchen 12 feet by 14 feet in their single-family home. The town requires that owners receive permission from the Planning Board to expand existing structures. Write a letter to the Planning Board, 142 Central Avenue, Tarrytown, NY 10532–2315 on behalf of the Tisches' request that a variance be granted for the construction to take place. Architectural plans will be enclosed with the letter.

CASE STUDY 5: INVITATION LETTER

Read Case Study 5; then compose the letter.

Case: Send letter from Susan Jackson, Expo Coordinator, to Mr. Joel Friedman, property manager of AMS Management Company, 14 Maiden Lane, Irvington, NY 10500.

The Tenth Annual Co-op and Condo Expo is scheduled for Thursday, October 15, at the New Yorker Hotel. This conference is a free one-day event for property managers and members of the real estate industry.

When you write to Mr. Friedman, mention to him that he will be interested in the following seminars: Enforcing House Rules, Safeguarding an Apartment's Value, and Money-Saving Ideas for a Building. All individuals attending the Expo will find booths that exhibit products and services for building maintenance.

In addition, property managers and real estate agents will receive free advice on building operations and legal matters when dealing with clients. To attend the Expo, individuals must preregister on the web site *http://www.expo.com* or call 1-800-543-1000.

Note: When composing the letter, list the three seminars in outline form.

CASE STUDY 6: E-MAIL MEMORANDUM

Read Case Study 6; then complete the e-mail memo.

Case: Your manager, Luis Ortega, asks you to send an e- mail to John Ling (ling@gazette.com), a real estate agent, to inform him of an article that appeared in the *Gazette Weekly*. This article featured the suburban town of Greenburg, the area where John Ling sells real estate. Mr. Ortega believes that John Ling will find the article informative since it describes in detail the positive aspects of the Greenburg community, its zoning restrictions, prices of area homes, and future development.

The subject of this e-mail is Gazette Weekly Article on Greenburg Community.

ASSESSMENT: CHAPTER 5

English Power

Directions: *Insert commas where required. If no comma is needed, write "C" (Correct) in the space provided.*

__________ 1. Our company provides professional property management and it offers individual attention to deal with specific situations.

__________ 2. The property has easy access to the downtown area and is located near the interstate highway.

__________ 3. The bank denied Mr. and Mrs. Jackson a home mortgage but they will reapply at another financial institution.

__________ 4. Charles Jones will speak at the real estate meeting on environmental concerns and he will hold a question-and-answer session following the presentation.

__________ 5. We plan to hold a public meeting to review the site plans and get reactions from the community.

__________ 6. The plan for rebuilding the inner-city neighborhood was approved but the funding will not be available until next year.

__________ 7. building was declared an historic site and that restricts its use for commercial purposes.

__________ 8. The firm will examine the building for asbestos and it will offer a plan for proper removal.

__________ 9. The public is concerned over the hazards of environmental contamination and is lobbying for stricter construction codes.

__________ 10. The large tract of undeveloped land will be sold to a developer or it will be subdivided into small lots for single houses.

Editing Power

Directions: *Edit, format, and proofread the following unarranged letter. Supply missing parts and paragraphs; key it correctly.*

MS ROSE DESANCTIS, MANAGING AGENT, ROCKLEDGE HOUSE, 22 ROCKLEDGE AVENUE, MEDFORD, MA 02155-3290

The Energy Fuel Group invite you to join 2,000 other members to receive top-quality oil-heating service for your building, and a guaranteed saving on fuel bills. We offer a budget plan that allows our members the convenience of prorating there oil payments equally throughout the year, and reducing huge bills during the winter months. You will receive an annual tune-up and cleaning of your furnace and we guarantee a 24-hour emergency service. You can apply for our plan by filing out the enclosed form. One of our representative will meet with you to discuss the energy needs of the building. Mark Kale, Manager

Writing Power

Directions: *Draft a form letter to customers who are delinquent in making mortgage payments on their homes for more than three months.*

This letter should indicate that the bank did send monthly reminders on overdue payments. In your letter, notify the customers that after three months, the bank takes legal action to pursue foreclosure of the property.

(Remember that this letter is sent at the final stage of the collection process.)

> "Diversity: the art of thinking independently together."
>
> —*Malcom S. Forbes*

6

Maintaining Healthy Complex Sentences

WHAT YOU WILL LEARN IN THIS CHAPTER

- To identify complex sentences
- To punctuate complex sentences
- To punctuate parenthetical expressions
- To develop a business vocabulary in **HEALTH CARE**
- To proofread and edit correspondence and reports
- To compose letters of refusal
- To compose a collection letter
- To compose a sales letter
- To write an e-mail memo
- To compose an informational memo report

Corporate Snapshot

PFIZER, INC., one of the world's leading health-care companies, discovers, develops, and markets medicines for humans and animals. Fundamentally, Pfizer's business is about saving, improving, and enhancing lives. It values the highest ethical standards to manufacture quality products both in prescription and over-the-counter drugs.

Pfizer is committed to being an exemplary corporate citizen. This is reflected in the company's extensive efforts to actively improve access to health care both in the United States and around the world, to care for the environment, and to encourage employee volunteerism in the community. Pfizer employees can be found offering their services in health organizations, hospitals, schools, and libraries. This firm is an example of a corporation that utilizes its size and resources to better the global community.

Source: Pfizer, Corporate Communications

ACTIVITY A LANGUAGE ARTS RECALL

Sentence Structure: Complex Sentence

A **complex** sentence consists of an **independent** clause and one or more **dependent** clauses that appear anywhere within the sentence. (A clause is a unit that includes a subject and a verb.)

An **independent** clause, as you reviewed in Chapter 5, can stand alone as a sentence because it has a subject and a predicate (verb and its modifiers) and expresses a complete thought.

Example: Good ***nutrition and exercise are*** two basic factors in maintaining a healthy body.

A **dependent** clause does not express a complete thought; therefore, it cannot stand alone.

Example: ***If individuals exercise and follow nutritious diets,*** they will maintain healthy bodies. (The dependent clause cannot stand alone because it does not express a complete thought.)

Comma Usage in Complex Sentences: Introductory Clauses

Rule 6.1: When there is an introductory clause at the beginning of a sentence before the independent clause, place a comma after it.

Example: ***As we discussed yesterday,*** a registration form will be mailed to you. (introductory clause)

Rule 6.2: A dependent clause at the end of a sentence in its usual grammatical position is usually not set off by a comma.

(When the clause is at the beginning of a sentence, it is an introductory clause.) Example: You may be interested in the health options ***although you are enrolled in the community insurance program.***

Although you are enrolled in the community insurance program, you may be interested in the health options.

Comma Usage: Introductory Phrases

A **phrase** consists of two or more words that do not contain a subject and a verb. It expands the meaning of the sentence.

Example: ***If available,*** our spring catalog will be sent to you.

Rule 6.3: When a phrase occurs at the beginning of a sentence before the independent clause, place a comma after it.

Example: ***At the beginning of the month,*** a seminar on medical insurance plans will be held. (introductory phrase)

Rule 6.4: When an introductory clause or phrase is the subject of a verb, do ***not*** place a comma after it.

Example: ***To secure adequate health insurance coverage*** is a necessity for individuals and families.

Comma Usage: Parenthetical Expressions

Parenthetical expressions are unnecessary words that do not add meaning to the sentence.

Rule 6.5: Use commas around parenthetical expressions (unnecessary words) that occur within a sentence or at the beginning of a sentence.

Example:

In our opinion, a greater number of people maintain a healthy diet and follow an exercise program.

Nutritionists, ***without a doubt,*** recommend a healthy diet and moderate exercise.

Rule 6.6: If a phrase at the end of a sentence gives additional but nonessential information, place a comma before the phrase.

Example: Individuals are interested in weight loss diets, ***such as low carb diets.***

Check for Understanding

Directions: Compare the sentences and answer the questions below.

Phrases and Clauses

1. If you belong to an HMO plan, you have a limited choice of doctors.
2. You have a limited choice of doctors if you belong to an HMO plan.
3. To control costs, we are increasing the deductible on medical insurance.
4. We are increasing the deductible on medical insurance to control costs.
5. To offer a national health plan is the senator's proposal.

Questions

a. What do sentences 1 and 3 have in common?

____

b. What do sentences 2 and 4 have in common?

____

c. How is the structure in sentences 1 and 3 different?

____

d. Why isn't there a comma after the phrase in sentence 5?

____

(Answers are given on pages 134)

Conclusion: Complete the following sentence.

Place a(n) _______ after a(n) _______ phrase or clause at the beginning of a sentence.

Parenthetical Expressions

6. Many corporations, for example, have set up optional health insurance plans.
7. Undoubtedly, the cost of medical insurance is rising.
8. You will realize, without a doubt, that our health benefits are superior.
9. Finally, the committee completed its findings.

Questions

e. What is similar about the phrases in sentences 6 and 8?

____

f. How is a parenthetical phrase within a sentence punctuated?

____

g. What is similar about the first word in sentences 7 and 9?

____

h. How is a parenthetical word at the beginning of a sentence punctuated?

____

(Answers are on page 134.)

Conclusion: Complete the following sentence.

When there is a(n) ________ word or phrase within a sentence, it must be set off by ________.

Answers to "Check for Understanding"

a. There is a comma after the introductory clause and phrase.
b. The phrase and clause appear in their usual grammatical order; therefore, no comma is required.
c. Sentence 1 has an introductory clause, and sentence 3 has an introductory phrase.
d. The introductory phrase is the subject of the sentence.
e. There is a parenthetical expression within each sentence.
f. There is a comma before and after the parenthetical phrase.
g. Each word is a parenthetical.
h. There is a comma after the parenthetical expression.

Building English Power: Introductory Clauses and Phrases and Parenthetical Expressions

Directions: Insert commas where required. If no comma is needed, write "C" (Correct) in the space provided.

________ 1. For many workers health care is a major expense. (6.3)

________ 2. If you wish to continue to participate in the insurance plan please return the signed form. (6.1)

________ 3. To include vegetables and fruits in your daily diet is recommended for good health. (6.4)

________ 4. Since your last bill premium rates have increased slightly. (6.3)

________ 5. You will receive supplemental insurance benefits if you are hospitalized. (6.2)

________ 6. More than ever you need adequate medical insurance. (6.5)

________ 7. We cannot process your claim however without a detailed statement. (6.5)

________ 8. In an effort to control costs we increased the deductible on the dental plan. (6.3)

________ 9. We have served one million members since we offered our plan. (6.2)

________ 10. To exercise regularly is important for physical fitness. (6.4)

________ 11. As you know a smoke-free workplace is endorsed by most Americans. (6.3)

_______________ 12. DWI laws without a doubt have become stricter in the past decade. (6.5)

_______________ 13. After the public debate Congress will address the issue of health insurance. (6.3)

_______________ 14. Within the last few years HMOs have become an alternative approach to medical treatment for patients. (6.3)

_______________ 15. Arthritis problems according to some doctors can be treated with a proper nutritional diet. (6.5)

ACTIVITY B BUILDING WORD POWER

The words below are commonly used in the health-care industry. Study the meaning of each word. From the list below, select the word or phrase that correctly completes each statement. Refer to the Personal Dictionary in the Appendix for the meaning of each word.

biomedical	network
cholesterol	nutrition
comprehensive	nutritionist
DWI	ophthalmologist
HMO	orthopedic
mammogram	outpatient
metabolism	transition

_______________ 1. The eye infection was treated by a(n) ________.

_______________ 2. Some people have high levels of blood ________, which may indicate an excessive amount of fat in the diet.

_______________ 3. A balanced, healthy diet is affected by sensible ________.

_______________ 4. Advanced technology has brought about ________ systems, which interconnect communications throughout the world.

_______________ 5. ________ groups, or health maintenance organizations, offer medical services and doctors to their subscribers.

_______________ 6. The football player consulted a(n) ________ doctor who specializes in bone injuries.

_______________ 7. The ________ from one stage of life to another can be stressful.

_______________ 8. ________ is the process by which the body converts food, oxygen, and other substances into tissues and energy.

_______________ 9. The ________ met with the patient to plan a low-cholesterol diet.

_______________ 10. The research team completed a(n) ________study that describes in detail the relationship between smoking and heart disease.

_______________ 11. ________ is a term that refers to high-tech medical equipment used in the replacement of parts of the human body.

_______________ 12. A(n) ________ department treats individuals who go to a hospital on a regular basis for treatment.

_______________ 13. Women are advised to have a(n) ________ every year to examine for breast cancer.

_______________ 14. ________ laws refer to those that penalize individuals who drive while intoxicated.

On the Alert

- affect (verb)—to influence

  Example: Diet ***affects*** an individual's health.

  effect (noun)—result

  Example: Certain foods have a positive ***effect*** on your health.

  effect (verb)—to bring about change

  Example: The new administration has ***effected*** changes in health coverage.

- assistance (noun)—aid

  Example: After the flood, the community received financial ***assistance*** from the government.

  assistants (plural noun)—people who assist

  Example: The physician's ***assistants*** administered the medical procedure.

- patient (adjective)—1. able to wait, showing self-control

  Example: The passenger was ***patient*** after the delayed flight.

  2. patience (noun)—the ability to remain calm

  Example: Nurses need a great deal of ***patience*** when dealing with a troubled patient.

  3. patients (plural noun)—individuals receiving medical treatment

  Example: The ***patients*** waited a long time at the clinic.

- to (preposition)—express direction

  Example: The free tickets ***to*** the seminar on nutrition were sent ***to*** your house in the mail.

  too (adverb)—1. more than enough

  Example: The patient is ***too*** sick to leave the hospital.—

  2. also

  Example: Mary's brother is attending the health fair, ***too.*** ("Too" is used as a parenthetical expression.)

- Use a hyphen between compound words that begin with "self."

  Example: self-esteem

- Hyphenate compound adjectives (one-thought modifiers) that precede nouns.

  Example: one-year study

- Omit the word *dollars* after the number when the dollar symbol ($) is used.

  Example: $200

ACTIVITY C INDUSTRY PERSPECTIVES: PHYSICAL FITNESS

Directions: Using appropriate revisions, proofread and key the report below. There are some built-in punctuation errors.

Rule Reference

6.3 In america today individuals care about physical fitness and good nutrition. The trend is toward a lifestyle that balances moderate exercise with a

6.1 sensible diet. As we know individuals who are physically fit are better able to cope with daily stress.

Doctors recommend a regular routine of physical activity for thirty minutes that is preformed three days a week. Doctors warn against strenuous exercise, which

Rule Reference	
	can be harmful. Sensible exercise, such as running and
6.5	walking can help raise your metabalism and control
	weight to.
	A good diet plan and regular exersise will help to
5.2	lower cholesterol levels, and reduce heart problems. As
6.1	recommended by nutritionists fruits, vegetables, and
	whole grains should be the basis for every diet. Diet
6.5	fads on the other hand are not the answer to good
	health. Individuals who follow these diets often regain
	lost weight. This failure to control weight can effect a
	person's self esteem. To develop a long-term fitness
6.4	plan, is the best way to maintain lifelong health.
6.4	To be physically and mentally fit is a goal of every
6.5	person. This goal however can only be achieved by a
	long-term commitment to exercise and nutritious eating.

Did you find these errors?
11 punctuation
1 hyphen
1 capitalization
3 typographical/spelling
2 word usage

ACTIVITY D EDITING POWER

Task A: Edit, format, and proofread the following form letter for introductory phrases and clauses as well as for parenthetical expressions and other errors. Use proofreaders' symbols to mark the errors, then key the letter correctly. Use "Dear Participant:" for the salutation.

Errors

- commas
- subject-verb agreement
- capitalization
- hyphen
- spelling
- typographical
- word usage
- number usage

As a result of a comprehensive one year study on medical insurance costs sections of our medical plan has been restructured.

In an effort to control future costs the following changes will become affective immediately:

1. Your deductable will increase to $200 for an individual and $350 dollars for a family.
2. There will be a five percent increase in the premuim rate.
3. You will need a second opinion for major surgery. If this procedure is not followed payment will be denied.

You may wish however to consider our monthly payment plan, if you find that an increase in premiums may be difficult to absorb. Please feel free to call our Customer service Department at (800) 336-1000, if you need assistants.

Sincerely yours,

Cynthia Washington
Director

Task B: Edit, format, and proofread the following unarranged memorandum for all errors. Supply missing parts and paragraphs; key the memorandum correctly.

This memorandum should be sent to the staff of the Department of Orthopedics from Dr. Luis Alvarez, Chief of Orthopedics. The subject is "Moving Procedures."

You will be happy to know that the new surgical wing of the hospital will be in full operation on September 1, 20___. This wing has the most advanced biomedical equipment and patience services. The move will take place on August 31 and the department will occupy the fifth floor of the building. The staff is asked to label all equiptment and supplies which needs to be transferred. Special boxes therefore will be provided to pack all your supplies and records. As of 5 p.m. on August 30 all computers will be shut down. Therefore be sure to have backup copies of all data stored in the computer. In your new quarters you will have a state-of-the-art telephone network system. These new telephone numbers however are not available now. Please be patient during this transition period and your corperation will be appreciated.

ri

ACTIVITY E WRITING PRACTICUM

Descriptive words can express messages distinctly and vividly. They can make a dull sentence come alive.

EXERCISE 1 REPLACING DESCRIPTIVE WORDS

Rewrite each sentence. Replace the general descriptive word (italicized) with a different, specific term.

Example: The lecture on controlling health costs was ***interesting.***

Rewrite: The lecture on controlling health costs was ***informative.***

1. The National Heart Association states that foods filled with saturated fats are *bad* for your health.

   Rewrite: ______________________________

   ____

2. Selecting a *good* medical plan can be one of the most important choices a family can make.

   Rewrite: ______________________________

   ____

3. Reading numerous health coverage brochures can be *boring.*

   Rewrite: ______________________________

   ____

4. The nurse who administered my blood test was *nice.*

   Rewrite: ______________________________

   ____

5. In addition to improving your health, a well-structured exercise plan can be *fun.*

   Rewrite: ______________________________

   ____

6. Filling out medical forms is a *dull* activity.

   Rewrite: ______________________________

   ____

7. Everyone hopes that medical reimbursement checks are issued *fast.*

   Rewrite: ____

   ____

8. The choice of a physician is an *important* factor when selecting a health plan.

   Rewrite: ____

   ____

9. My yearly physical revealed that I was in *fine* health.

   Rewrite: ____

   ____

10. Recuperating after a surgical procedure can be a *slow* process.

   Rewrite: ____

   ____

EXERCISE 2 CHOOSING POSITIVE OPENING SENTENCES IN LETTERS OF ADJUSTMENT AND REFUSAL

Select the statement that best expresses opening sentences with a positive tone and give the reason for your choice. Choose Option 1 or 2 in each group. Check the "Writing Pointers" before completing this exercise.

A. Topic: Refusing an Invitation

1. Your organization, Society for Protecting the Environment, has honored me with an invitation to address its annual meeting. I look forward to your professional and exciting meetings; however, I am unable to accept this invitation.
2. I received your invitation to address the annual meeting of the Society for Protecting the Environment. I am sorry that I cannot accept this invitation.

   Option 1. _______ 2. _______

   Reason: ____

B. Topic: Refusing an Invitation

1. Thank you for the invitation to be a moderator for the one-day workshop. I am sorry to say that I cannot accept this invitation.
2. I was pleased to receive your request to give a one-day workshop to your employees on critical decision making. I always enjoy leading a workshop on this important topic; however, I have a previous commitment.

Option 1. _______ 2. _______

Reason: ____

C. Topic: Refusing a Claim

1. We are sorry that we cannot process your prescription claim as submitted. We need additional information.
2. We have received your prescription claim for reimbursement. However, we are unable to process this claim without further explanation of the charges.

Option 1. _______ 2. _______

Reason: ____

D. Topic: Refusing a Claim

1. We are sorry that we cannot process your claim for the routine eye examination performed by your ophthalmologist. Unfortunately, this procedure is not covered in your present medical policy.
2. You have submitted a claim for a routine eye examination performed by your ophthalmologist. This procedure is not covered in your policy.

Option 1. _______ 2. _______

Reason: ____

E. Topic: Refusing an Applicant

1. We appreciate your interest in enrolling in our comprehensive health insurance plan. However, your preexisting heart problem, unfortunately, makes you ineligible for our plan.
2. I am sorry we cannot offer you our comprehensive health insurance plan since you have a preexisting heart problem.

Option 1. _______ 2. _______

Reason: ____

F. Topic: Refusing an Applicant

1. You missed the deadline for our special rate for catastrophic health insurance. We can offer you the plan at the enclosed rates.
2. Thank you for expressing an interest in our catastrophic health insurance plan. We are sorry that the deadline for the special rates has elapsed. We have enclosed our present rates for coverage that is available.

Option 1. _______ 2. _______

Reason: ____

G. Topic: Requesting an Adjustment

1. I have received a reimbursement check for $75 for my recent foot examination charges. I believe that this amount is insufficient payment for the $200 fee that I submitted to you.
2. I received your reimbursement check of $75 for a recent bill submitted to you for foot examination charges. My records indicate that I may be entitled to a larger refund.

Option 1. _______ 2. _______

Reason: ____

H. Topic: Requesting an Adjustment

1. I cannot understand why I continue to be billed for an electrocardiogram test I took six months ago. This bill was paid by check at your office immediately after the test.
2. Your office has repeatedly billed me for an electrocardiogram test taken six months ago. Your records will show that this bill was paid immediately after the test in your office.

Option 1. _______ 2. _______

Reason: ____

Letters of Refusal

Letters of refusal must be written with tact so that the firm's goodwill will not be adversely affected. Never start the letter with bad news. With a friendly, pleasant opening, the reader is in a better frame of mind to accept a decision to refuse a request. Offer a reasonable explanation to support your answer. The "No" can be implied. The closing should be a positive statement indicating that the business relationship will continue.

Model Letter of Refusal

Date

MS LUCY BRITO
1042 MAPLE DRIVE
SACRAMENTO CA 95680-7391

Dear Ms. Brito:

Positive buffer statement

Thank you for your interest in enrolling in our Basic Health Plan. We would like to have you as one of our participants.

Reasonable explanation for refusal

After serious review of your medical background, we regret that we cannot accept you as one of our subscribers. A prerequisite for membership is that an applicant has not had a heart problem for the past five years. Your records indicate that you had a triple bypass last year.

Offer suggestions

There are other medical plans that focus on cases similar to yours. Unfortunately, we do not have such a plan. I suggest that you discuss this matter with your doctor or local medical facility.

Apology for non-acceptance

Goodwill statement

We are sorry that you cannot be a participant in our plan, but we hope that we can be of service to you in the future.

Sincerely yours,

Henry Sainville
Director

ri

Writing Pointers

Letter of Refusal

- Use an opening sentence as a buffer statement introducing the situational problem.
- Use a positive, upbeat approach; start a letter with a statement of thanks.
- Be direct but polite.
- Offer reasonable explanation.
- Offer alternative course of action and assistance.
- Apologize for inconvenience.
- Close with a goodwill statement.

ACTIVITY F WRITING POWER

CASE STUDY 1: REWRITE OF POORLY WRITTEN LETTER OF REFUSAL

After reading the poorly written letter sent to Alberto Genao, an individual who inquired about joining your medical group, respond to the questions that follow to help you organize your ideas before rewriting this letter. Then draft the letter. Send the letter to Mr. Genao, 21 Thornwood Drive, San Diego, CA 92100–2324.

Case

Dear Mr. Genao:

We would like to thank you for the interest in wanting to join our medical group.

Unfortunately, you are ineligible to join our group plan. Your company does not belong to our group. We allow only employees of companies that are members to participate.

Ask your human resources representative to suggest a medical group.

Questions

1. How would you rewrite the opening sentence to set a tone of appreciation and courtesy?

   ____

   ____

2. Following the opening statement, how would you convey the message of refusal?

   ____

   ____

3. What other statements should you include in this letter of refusal?

   ____

   ____

CASE STUDY 2: LETTER OF REFUSAL OF A CLAIM

After reading this case, respond to the questions that follow, which will help you organize your ideas. Then draft the letter to MS LORI YOUNG, 30 EAST LANE, DENVER, CO 80215–3460.

Case: Ms. Lori Young submitted a claim to her insurance company for her $200 dental expenses. The company reviewed it and did not accept the claim. The claims supervisor, Suzanne Martel, must notify Ms. Young of the claims refusal. Ms. Martel must inform Ms. Young that she should resubmit the claim with a detailed explanation of the dental procedure for a reevaluation of the claim. If Ms. Young has any questions, she can call the Claims Department office at 202-667-3000.

Organize Your Ideas

1. What is the most important point that Ms. Martel needs to make to Ms. Young?

___

___

___

2. What would you include in the opening sentence?

___

___

___

3. What specific details would you include in the body of the letter?

___

___

4. What should Ms. Martel say in her closing statement to maintain goodwill?

___

___

___

CASE STUDY 3: COLLECTION LETTER

Read Case Study 3, then develop an outline before composing the letter.

Case: A patient, Mr. Gerard Coleman, received six eye treatments from Dr. Larry Cutler. Mr. Coleman was informed before he started the treatments that his insurance would not pay for the procedure. The patient was willing to pay for the total cost of $1,200. The account is four months past due, and no payments have been made.

Write a letter on behalf of Dr. Cutler requesting payment.

CASE STUDY 4: REFUSAL OF AN INVITATION

Read Case Study 4; develop an outline to guide you in writing, then compose the letter.

Case: Your employer, Dr. Carolyn O'Connell, has been invited to speak at a regional medical conference in Kansas City on May 15, 20xx. The topic is "Stress in the Workplace." Dr. O'Connell is unable to accept the invitation because she will not be returning to the United States before the date of the conference. She is presently on a special assignment in South America. As administrative assistant to Dr. O'Connell, draft a letter to send her regrets and inform the group that she will be available at another time. Address the letter to Dr. Ivan Ruiz at County Medical Center, 344 Eastern Parkway, Kansas City, KS 64100–4301.

CASE STUDY 5: INFORMATION LETTER

Read Case Study 5; then compose the letter.

Case: The Americana Health Insurance Company is introducing a new prescription drug discount plan for its members. It is good news because the rates will be 15 percent lower than drug prices at the pharmacy.

It is easy to order the prescription drugs. Initially, the member sends in the doctor's prescription, and afterwards the individual simply calls in for refills. The order is mailed within a week. There are no shipping charges. There is overnight delivery for emergencies. With a generic drug, the savings are greater. The charge is billed directly to the credit card account.

There is also a health product catalog that lists many over-the-counter medications and beauty products at savings of 25 percent off listed prices.

Members can receive an information kit by returning the attached card. There is no obligation.

Send a form letter; saluation should read Dear Member. The letter is from Alicia White, Director of Sales.

Address Book

When the same e-mail message is to be forwarded to a group of individuals, a mailing group of names is created in the address book. When you send the e-mail message, you place the name of the mailing group in the **To** block. The message will be automatically delivered to all the names listed in the mailing group.

CASE STUDY 6: E-MAIL MEMORANDUM

Read Case Study 6; then complete the e-mail memo.

Case: Henry Kinnel, chief executive officer of Beard Pharmaceutical, is writing an e-mail to the staff announcing the approval of the drug Arpsin by the U.S. Food and Drug Administration. He congratulates the Research and Development Department for its work on developing the drug. Arpsin was tested for two years. The drug greatly improved the survival rate of heart attack victims. Mention that doctors will welcome this breakthrough. Arpsin will be available within a few months. Mr. Kinnel thanks everyone for their support in making the company a leader in its field.

Write the e-mail memo to be sent to all the staff members of the company. (In an office, this e-mail would be sent through the mailing group list in the address book.)

Include Mr. Kinnel's name and title in the closing.

The subject of this e-mail is Approval of New Drug Arpsin.

ACTIVITY G REPORT WRITING POWER: MEMO REPORTS

The ability to write reports is an important skill to master to communicate effectively in business. The two types of report writing are formal and informal.

The formal report, which is sent outside the company, is usually a lengthy, well-structured, and documented report. The informal report, both letter and memo format, is used widely to disseminate information. Within the organization, the memo report is the accepted format for communicating with employees. To summarize a problem, to present a rationale for a project, or to recommend specific practices, the one-page memo is accepted as the most expedient type of report.

In the next six chapters, Section G will focus on sharpening the skills of business report writing.

Information on Memo Reports

Functions

- To present relevant facts and information to help staff make decisions
- To provide analytical data
- To make recommendations based on factual information
- To report on the progress of a project

Types

- Status or progress reports to keep staff informed on the status of special projects
- Periodic reports prepared at set intervals to keep staff informed about company business (example: sales and financial reports)
- Informational reports to give facts, trends, and statistics to staff for decision making

Writing Pointers—Memo Reports

- Identify the purpose of the report.
- Write a clear, accurate introductory statement explaining the purpose of the report.
- Know your audience who will read the report.
- Write a draft and then edit and revise for correct English, grammar, and organization.
- Write in an informal manner; use of first person (I, we) is acceptable.
- Use memorandum format.
- Provide sideheadings for focus.
- State the purpose of the memo report in the subject line.

Model Informational Memo Report to Implement Change

TO: Staff Supervisors

FROM: Jane DeSanctis, Vice President
Human Resources

DATE: January 10, 2005

Purpose identified

SUBJECT: CHANGES IN EMPLOYEE PRACTICES AND PROCEDURES

Explains need for change

As newly appointed vice president of human resources, one of my priorities has been to review company policies and practices in dealing with our home health care personnel. We are very much aware that the employee turnover rate in the industry is high. However, at Wellcare Agency, the rate is above the norm. A firm's workforce is its greatest asset. High employee turnover and poor performance affect our business through increased costs and loss of clients.

Use of personal pronoun

I have met with each of you personally to discuss ways to improve employee relations and have sent a questionnaire to all home health care personnel. The responses revealed some disturbing facts; however, they also offered constructive suggestions to improve the working climate within the organization.

Side heading for focus

FINDINGS

Presents facts

During the past year, our turnover rate was 40 percent, and a poor working climate was the main reason for individuals leaving the agency. Responses from our present employees highlighted the following reasons for job dissatisfaction: inadequate on-the-job training, poor internal communications, lack of self-worth, and little recognition by management for outstanding work.

Provides recommendations based on findings

RECOMMENDATIONS

Based on the findings, the following changes in practices and procedures will be implemented immediately:

- Formation of a task force to review and improve on-the-job training

Clarity of reading

- Monthly meetings of supervisors with their employee teams to interact and communicate
- Involvement of personnel in decision making (example: updating home health care practices)
- Willingness of management to listen to the rank-and-file who are on site every day
- Periodic visits by supervisors to assess the care given to patients in their homes
- Implementation of a program to reward outstanding employees

It is through a cooperative effort that we shall be able to energize our employees to make Wellcare Agency a great place to work and to do business.

CASE STUDY 1

After reading the following case, respond to the questions in the section that follows in "Organize Your Ideas," then compose the memo report.

Case: You are working in the Community Relations Department of your local hospital. The neighborhood surrounding the hospital has a multicultural population, and many of the residents are recent immigrants who are not familiar with our health services.

As a member of the Community Task Force, you have been involved in a recent study that shows a need to offer programs in English and other languages for expectant parents. These programs would include courses on infant care and parenting skills. The need is supported by data that shows that 50 percent of the respondents spoke English as a second language. Of the 500 families who responded, 30 percent of them have been living in the community less than three years and are unaware of the services provided by the hospital.

In a memo report to Dr. Louis Farnsworth, director of the hospital, discuss the findings of the study, the educational programs recommended by the Task Force, and the implementation of courses. The recommendations include offering monthly classes to discuss topics of interest for expectant parents; conducting tours of the state-of-the-art maternity, nursery, and delivery units; forming a support group of knowledgeable parents to teach parenting skills; and holding weekly classes on infant care. These programs should be bilingual to accommodate the community. At present, a bilingual evening course on parenting skills and a weekend class on infant care have been started.

Remind the hospital administration that it is important for the hospital to reach out using a preventive approach to avoid health problems later. Also, invite Dr. Farnsworth to meet with the Task Force.

Organize Your Ideas

To help you write an effective report, respond to the following questions before composing it. The questions are organized in a logical sequence; therefore, as you write the report, use your responses to the questions as a reference. When completed, your response list, in effect, will be the outline you should follow step by step during the writing process. This section reviews important points that need to be considered when writing this report.

1. What is the purpose of this memo report?

   ____

2. Which of the following is the most appropriate subject line?
   a. Proposed Health Education Program
   b. Community Task Force Study
   c. Implementation of the Community Task Force's Recommendation
3. What will be the sideheading for your opening?

   ____

4. What information do you plan to give the reader in the introductory paragraph?

   ____

5. Which of the following is the most appropriate opening sentence of the memo report?
   a. I am reporting on behalf of the Community Task Force that the study on the health needs of our residents is completed.
   b. The study of the Task Force has completed its study of the health needs of the residents of our community.
   c. The Community Task Force has completed its study of the health needs of the residents of our community.

6. What other information would you include in your opening paragraph?

____

7. What is the sideheading for the next section of your report?

____

8. What is the best way to present the recommendations that are being implemented?

____

9. What key words will your recommendation include? (Put the key words in order to form an outline.)

____

10. What would you say in the closing section of the report?

____

CASE STUDY 2: MEMO REPORT

Community General Hospital serves a 20-square-mile area with a population of 25,000 residents. Ninety percent of the families are middle income, and the remaining 10 percent are low income.

The Community General Hospital's emergency room is always crowded and often there is a two-hour waiting period. To alleviate this overcrowding condition, a pilot project was started by opening a neighborhood clinic on the opposite side of town. The clinic has been in operation for one year. It is open seven days a week from 8 a.m. until 11 p.m. and is staffed by physicians from the hospital, part-time specialists in child care and senior care, nurses, and technicians.

Over the past year, treatment at the clinic has included emergency and routine situations such as sprains, colds, fevers, chest pains, and x-rays. The clinic has been well received by the community and treats about 250 patients a week. There is also a decrease in attendance at the emergency room of Community General Hospital.

Write a report to the hospital's Board of Directors informing them of the success of the pilot program. Suggest expansion of the clinic to serve that section of the city. Important services that could be added include pediatrics and family care.

ASSESSMENT: CHAPTER 6

English Power

Directions: *In the following sentences, insert correct punctuation where needed.*

1. If you have a high level of cholesterol you should try to reduce fat in your diet.
2. The use of steroids moreover is dangerous to an individual's health.
3. As you know your doctor can prescribe generic-brand drugs.
4. Yearly physical examinations without a doubt lead to the early detection of serious illnesses.
5. Some doctors agree to reduced fees if patients have a limited income.
6. People who are enrolled in certain health plans as you may know are required to use a participating doctor.
7. As we discussed the nonsmoking laws will affect the workplace.
8. Most medical plans however no longer cover laboratory tests in full.
9. If the new health plan is passed more people will be eligible for benefits.
10. Contrary to past public opinion statistics show that heart disease is on the rise among women.

Editing Power

Directions: *Edit, format, and proofread the following unarranged letter for errors; key the letter correctly.*

Ms. Anne Barnes, 106 Ocala Drive, Ft. Meyers, FL

33919-1500

Dear Ms. Barnes: For a great many people a health-conscious lifestye has become a challenging task. If you are looking for an interesting guide to a nutritious diet we recommend a new book called Healthy Foods. This most informative manual has been written by Dr. Edward Jaffe, a well-known nutritionist. In the book Doctor Jaffe recommend foods that help prevent some diseases. Studies without a doubt shows that certain fruits, vegetables, and grains are rich in vitamins and minerals which have a positive affect in fighting illnesses. The author also offers valuable advise on sensible dieting. There is a section on easy-to-prepare recipes which use foods that have high nutritional value. We invite you to preview Healthy Foods for 30 days by mailing the enclosed, self addressed reply card. If you decide to keep this delightful book just send us a check for $25. Otherwise return the book to us at no charge to you.

Sincerely, Charles Gibson, Sales Manager

Writing Power

Directions: *Read the following case; then draft a letter to Health Incorporated, 2105 King Boulevard, Tuscon, AZ 85725-7500.*

Your insurance company, Health Incorporated, reimbursed you only $50 for a bill submitted to them for your mammogram. The charge for the test was $150. You are requesting that they review this claim since you have met the minimum requirement of $200 from a previous bill this year. You feel that you are entitled to a greater reimbursement.

> "Talent wins games, but teamwork and intelligence win(s) championships."
>
> —*Michael Jordan*

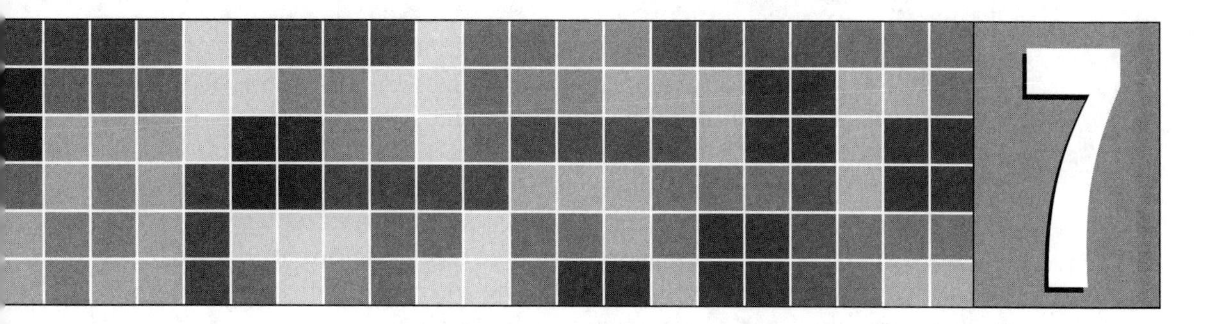

Retailing Commas in Words, Phrases, and Clauses

WHAT YOU WILL LEARN IN THIS CHAPTER

- To apply commas to nonrestrictive clauses and phrases
- To punctuate a series of items
- To apply commas to words and phrases in apposition
- To punctuate consecutive adjectives
- To punctuate items in dates
- To develop a business vocabulary in **RETAIL SALES**
- To proofread and edit correspondence and reports
- To compose a reply to a customer's letter of complaint
- To compose a reply to a letter of inquiry
- To compose a promotional letter
- To compose a periodic memo report
- To compose an informational letter
- To write an e-mail memo

Corporate Snapshot

TARGET, an upscale discount chain, is recognized as one of the nation's top retailers. It provides quality merchandise at attractive prices in customer-friendly stores.

Target's commitment to respect for individuality and importance of diversity is its formula for success. Its employees represent a wide spectrum of race, gender, age, ethnic heritage, religious affiliation, education, and experience. Respect for the individual ensures a tolerant workforce that works as a team to achieve results.

Target consistently receives awards for the employment opportunities it provides for minorities and women, many of whom are in senior leadership positions. Target was named as one of the best companies for working mothers in recognition of its policy of helping parents balance work and home life. Target has a reputation as a good place to work and shop.

Source: *http://target.com/target corp_group/company/index.jhtml*

ACTIVITY A LANGUAGE ARTS RECALL

Comma Usage

Commas that are used within a sentence show relationships between words, phrases, or clauses.

Series of Words, Phrases, and Clauses

Rule 7.1: Use a comma after each item in a series of words, phrases, or clauses, including the item before the conjunction "***and, or, nor.***"

Example: We stock ***accessories, luggage, and stationery.***

Nonrestrictive and Restrictive Clauses

Rule 7.2: Use commas around nonrestrictive clauses and phrases (clauses that are not essential to the meaning of the sentence).

Example: Your catalog order, ***which was received today,*** will be shipped immediately.

Rule 7.3: Do not use commas around restrictive clauses (clauses that are essential to the meaning of the sentence).

Example: The customer ***who returned the damaged item*** was given a refund.

Rule 7.4: Do not use a comma before the word *that* when it introduces a clause that is essential to the meaning.

Example: The one-day sale ***that was scheduled for January 4*** is canceled until further notice.

Apposition

Rule 7.5: Use commas to separate words or phrases in apposition (when words within commas explain the preceding noun or pronoun).

Example: Her first book, ***Steps to Successful Selling,*** was just published.

Rule 7.6: Do not use a comma when a word or phrase in apposition is essential to identify the preceding noun.

Example: The new book ***How to Shop*** is offered at a discount to customers.

Consecutive Adjectives

Rule 7.7: Use a comma to separate two or more consecutive adjectives that modify a noun. To test for a comma, mentally insert the word *and* between the two adjectives or reverse the order of the adjectives.

Example: Soft, silky fabric was used in our collection of scarves. (Test: Soft and silky fabric. . . .)

Dates

Rule 7.8: Use commas to set off items in dates. When a year is used with other items in a date, place a comma after the year.

Example: ***March 1, 2005,*** is the date we scheduled for the flea market sale.

Omit commas when only month and year are given.

Example: The May 2005 issue is an anniversary edition.

Cities and States

Rule 7.9: Use a comma between the name of a city and the name of a state. Also, place a comma after the name of the state when it is preceded by the name of a city and is not followed by final punctuation.

Example: We will be visiting ***Charleston, West Virginia,*** and ***Asheville, North Carolina.***

Check for Understanding

Directions: Compare the sentences below and answer the questions that follow.

Nonrestrictive and restrictive clauses

1. The customer who was dissatisfied with the merchandise complained to the manager.
2. The customer, who had purchased the item on sale, returned it for a refund.
3. Our upscale line of clothing that has shown a net loss will be discontinued immediately.
4. Our upscale line of clothing, which is popular with our customers, will be expanded.

Questions

a. What do sentences 1 and 3 have in common?

____

b. What do sentences 2 and 4 have in common?

____

c. Why do sentences 1 and 3 have no internal punctuation?

____

d. Why do sentences 2 and 4 have internal punctuation?

____

(Answers are given on page 163.)

Conclusion: Complete the following sentence.

Use no ________ around restrictive clauses; use ________ around nonrestrictive clauses.

Apposition

5. We have enclosed our shopping credit card, Royalcharge, for your convenience.
6. The new credit card Royalcharge will offer shoppers many added features.

Questions

e. What is the difference in punctuation between sentences 5 and 6?

____

f. Why is the word *Royalcharge* set off by commas in sentence 5?

____

g. Why are there no commas in sentence 6?

____

(Answers are given below.)

Conclusion: Complete the following sentence.

When a word or phrase is necessary to identify the preceding noun, do ________ use a comma. If the word or phrase describes the preceding noun, it is in ________; therefore, ________ need to be inserted.

Answers to "Check for Understanding"

a. There are no commas in the sentences.
b. There are commas in the sentences.
c. There is a restrictive dependent clause in each sentence.
d. There is a nonrestrictive dependent clause in each sentence.
e. Sentence 5 has commas; sentence 6 has no commas.
f. *Royalcharge* is in apposition and explains the preceding noun ("credit card").
g. The word *Royalcharge* is necessary to identify the preceding noun (it is the name of the new credit card).

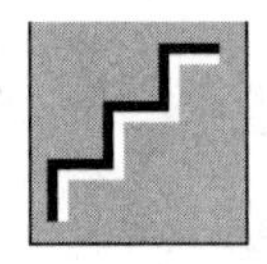

Building English Power: Comma Use in Series, Consecutive Adjectives, Nonrestrictive/Restrictive Clauses, Apposition, Dates, and Cities and States

Directions: Insert commas where required. If no comma is needed, write "C" (Correct) in the space provided.

_______________ 1. You a valued customer are eligible for our money-saving bonus plan. (7.5)

_______________ 2. On May 1 20xx our popular well-known rug sale will take place. (7.8, 7.7)

_______________ 3. Price Cutter a national chain will open a store in Huntington New York a thriving suburban community. (7.5, 7.9)

_______________ 4. Designer clothing which appeals to today's shopper can be purchased at discount stores. (7.2)

_______________ 5. Appliances carry a manufacturer's warranty that offers consumers protection against product defects. (7.4)

_______________ 6. Shoppers who look for affordable quality are joining warehouse price clubs. (7.3)

_______________ 7. Our electronics store which is located in the mall is offering a 10 percent discount on VCRs camcorders and CD players. (7.2, 7.1)

_______________ 8. As a valued respected charge customer you are invited to our annual two-day sale which will be held this weekend. (7.7, 6.1, 7.3)

_______________ 9. If you are interested in discount stores Chet Edwards' latest guide *The Popular Discount Malls* is now available at your bookstore. (6.1, 7.5)

_______________ 10. The seminar speaker discussed today's competitive market consumer preferences and new marketing trends. (7.1)

_______________ 11. The new upscale mall that attracted huge crowds on the weekend will extend its shopping hours. (7.7, 7.4)

_______________ 12. The popular home-furnishing store offers free shipping a convenient layaway plan and low credit rates. (7.1)

_______________ 13. The firm has scheduled Friday July 31 for its annual physical inventory. (7.8)

_______________ 14. Our spring catalog which was recently mailed to you illustrates the latest women's fashions. (7.2)

_______________ 15. The best seller *Repackaging Your Sales Approach* sells for $19.95 plus shipping. (7.6)

ACTIVITY B BUILDING WORD POWER

The words below are commonly used in the retail sales industry. From this list, select the word or phrase that correctly completes each statement. Refer to the Personal Dictionary, if necessary.

backlogged	physical inventory
boutiques	privileged
electronic malls	redeemed
franchises	upscale
innovative	upsurge
megamalls	warehouse clubs
outlet malls	warranty

_______________ 1. Shoppers find that ________ are good places to purchase large quantities of popular products at low prices.

_______________ 2. High-styled clothing is sold in small, specialized stores known as ________.

_______________ 3. Through the Internet, shoppers will be using ________ to make their purchases.

_______________ 4. ________ are huge shopping sites that have expanded the variety of retail stores to include restaurants, movies, and entertainment.

_______________ 5. When economic conditions are poor, there is a(n) ________ of unemployment.

_______________ 6. Airlines offer bonus points for traveling that can be ________ for free trips.

_______________ 7. ________ people can often find different, creative ways to design projects and solve problems.

_______________ 8. The popular children's video has sold out, and production has been ________ for a month; therefore, shipment of the video will be delayed.

_______________ 9. Well-known personalities from the fields of sports and entertainment are ________ individuals who can gain access to social events.

_______________ 10. Clothing manufacturers often have stores in ________ to sell their overstocked and discontinued line of merchandise at lower prices.

_______________ 11. The shopper was interested in visiting the ________ mall that features boutique stores and higher-priced department stores.

________ 12. Retail stores conduct a(n) ________ each year to manually count merchandise for control purposes. (In retailing, the word *inventory* means "merchandise.")

________ 13. Most appliances carry a manufacturer's ________ that guarantees against defective parts of a product.

________ 14. Many retail firms offer ________ to buyers, which gives them the right to use the company name and product while they adhere to specific company policies and conditions.

On the Alert

- build (verb)—to construct

  Example: The Town Planning Board voted to ***build*** a new mall.

  built (verb, past tense)—constructed

  Example: This historic home was ***built*** in the 1900s.

- on-site—on location (e.g., ***on-site*** hotel; compound adjective modifying the noun)

  Example: The fire department carried out an ***on-site*** inspection of the building.

- plural of numbers—Add "s" without an apostrophe.

  Example: During the ***1990s,*** the Internet became an alternative way of personal shopping.

- percent—Express percentages in figures except at the beginning of a sentence; spell out the word *percent.*

  Example: 10 percent

- punctuation and question marks—Periods and commas are always inside the quotation mark.

  Example: I read that his latest article, "Selling Tips," is available.

- books—Titles of books are generally italicized. All caps may be used to highlight a title.

  Example: *Repackaging Your Sales Approach*

  REPACKAGING YOUR SALES APPROACH

- Internet—Capitalize I in this widely used technology term.
- Decades and centuries—write as words and do not capitalize.

ACTIVITY C INDUSTRY PERSPECTIVES: RETAILING TRENDS

Directions: Using appropriate revision marks, proofread and key the report below. Refer to the Appendix for a model report.

Rule Reference	CURRENT RETAILING TRENDS
	The retailing industry has changed dramatically
	during the passed decade. Consumer demands for quality
	merchandise at competitive prices effect the way
	retailers sell their products. To meet the needs of
	specific consumer markets megamalls warehouse clubs and
6.4, 7.1	specialty boutique shops have become popular.
3.1, 7.2	The megamall which is especially popular with
2.1	families offer the consumer a combination of retail and
7.1	department stores fast-food eateries and entertainment
1.4	establishments. The mall has become the "Main Street of
	america." Also, the discount stores of the 1980s have
	been replaced by discount malls where outlet stores of
2.1	famous brand names offers customers merchandise at 20 to
	50% off the regular prices.
7.2	Warehouse clubs which are membership-only discount
	stores also are targeted for family-oriented shopping.
2.1	These supercenters offers a variety of bulk merchandise
6.3	at low prices. At the other end of the marketing
	spectrum specialty boutiques are favored by shoppers who
	believe that brand names reflect status symbols.

TV and the Internet have become powerful forces in

1.7 changing the shopping habits of americans. Individuals,

7.3 who have limited time, use the TV shopping channel to

make their purchases. The Internet with it's vast source

of information has been attracting retailers to sell

there merchandise through electronic malls.

6.3 In the 21st century retailing will not be "business

7.7 as usual" but an exciting innovative industry.

Did you find these errors?
12 punctuation
2 capitalization
3 word usage
3 subject-verb agreement
1 number usage
1 symbol usage

ACTIVITY D EDITING POWER

Task A: Edit, format, and proofread the following unarranged form letter. Use proofreaders' symbols to mark the errors, then key the letter correctly. Remember to insert paragraphs.

Errors
- commas
- verb tense
- word usage
- typographical

Dear Customer: We value you as a customer, and want to

show our sincere appreciation for your support and

loyalty. A special weekend has been reserve for our

preferred customers. This weekend, June 15 16 and 17 you

are invited to take 10% off our entire line of

merchandise which includes jewelry clothing and home

furnishings. To get this extra 10 percent of the ticketed price present this letter and use your Albertson's charge card when making you purchase. We are introducing a new service for our busy customers. The Shopper's Cafe will serve salads sandwiches and gourmet coffees throughout the day. We look forward to serving you. Sincerely yours, Sam Ross, Vice President

---

Task B: Edit, format, and proofread the following rough-draft letter for all errors; key it correctly.

Address this letter to Ms. Sylvia Atkins, 11 Secor Drive, Memphis, TN 38111-0700. The letter is from Thomas Butler, sales representative.

I invite you to take advantage of our convenient home-buying service Value Shopping. As a busy individual you know your time is precious. With our plan you do not face crowded stores long lines and limited stock. Our catalog which is published twice a year advertizes our fine line of fashion clothing. Our order forms are simple to to complete and we have an easy payment plan. With every order you receive bonus shopping points which can be redeemed toward future purchases. Please read this enclosed mailing and become a priviledged customer. We look forward to providing you with efficient reliable service.

ACTIVITY E WRITING PRACTICUM

Parallelism

Sentences should be easy to read and understand clearly. One way to ensure this is to express similar words in the same grammatical form to communicate corresponding thoughts.

Examples:

- pairing a noun with a noun
- pairing a verb with a verb
- pairing a phrase with a phrase
- pairing an infinitive with an infinitive ("to" + verb)
- pairing a clause with a clause

This construction is known as ***parallelism,*** which improves readability and enhances clarity.

Examples of Parallel Construction

Incorrect:

- The management team was responsible for ***company strategy, finance,*** and ***directing*** the administrative team. (noun forms mixed with verb form)

Correct:

- The management team was responsible for ***company strategy, finance,*** and ***administration.*** (noun form)

Incorrect:

- This month's window display was ***chic, creative,*** and ***appealed*** to consumers. (adjectives mixed with verb form)

Correct:

- This month's window display was ***chic, creative,*** and ***appealing.*** (adjective form)

Incorrect:

- Advertising is an important tool that retailers use to ***attract*** the attention of the public, ***influence*** a potential shopper, and ***for arousing*** the consumer's curiosity. (mixed verb forms)

Correct:

- Advertising is an important tool that retailers use to ***attract*** the attention of the public, ***influence*** a potential shopper, and ***arouse*** the consumer's curiosity. (infinitive form)

Incorrect:

- In the retailing seminar, I learned about fashion ***merchandise,*** customer ***service,*** and ***motivating*** employees. (noun forms mixed with verb form)

Correct:

- In the retailing seminar, I learned about fashion ***merchandise,*** customer ***service,*** and employee ***motivation.*** (noun form)

Directions: Revise the following sentences so that they are balanced and so that the parallel ideas are written in the same grammatical form. Occasionally, a word or phrase may have to be added or deleted in the sentence.

1. The goals of the manufacturer are to mass produce custom-made clothing and selling directly to the consumer. (infinitive form)

   ____

   ____

2. The antiquated marketing system has been updated by using computer technology and by employees being sent for training. (verb form)

   ____

   ____

3. In marketing, computers are used for product inventory control, to record daily sales, and to get customer feedback. (noun form)

   ____

   ____

4. Suburban malls have replaced downtown shopping areas because they are centralized, diversified, and parking is accessible. (adjective form)

   ____

   ____

5. Our entire line of merchandise, which includes jewelry, furniture, and introduces new fall fashions, is on sale. (noun form)

6. This month's sales of winter clothes will be a huge success because of two-day sales and 20 percent off retail price will be given. (noun form)

7. Come to the selling course prepared to take notes, listen to the guest speaker, and asking some questions. (infinitive form)

8. Your mail-order catalog was received yesterday and my order for furniture sent to the purchasing department today. (verb form)

9. Good salespersons should have knowledge and familiarity with their stock. (preposition form)

10. The department store, specializing in sporting goods, also features designer clothing, equipment to exercise, and has books for traveling. (noun form)

11. The mall has increased its budget for hiring security personnel and to prevent shoplifting. (verb form)

12. The company sent out direct mailers advertising the sale and to promote new merchandise. (infinitive form)

____

____

13. When the shopping mall was designed, the architect considered mall motif, laying out of the stores, and accessibility to roads. (noun form)

____

____

14. The shopper purchased groceries at the supermarket, at the drug store bought cosmetics, and bought a sweater at the boutique. (phrase form)

____

____

15. Service, being dependable, and having product value are the major goals of our firm. (noun form)

____

____

Writing Pointers

Reply to a Customer Complaint

- Respond immediately to a complaint.
- Begin with a pleasant buffer statement.
- Offer a logical explanation for the action taken.
- Offer to correct the problem, if possible.
- Offer alternative solutions where applicable.
- Be sympathetic but firm.
- Respond positively to maintain goodwill.
- Use a friendly closing.

Letter of Promotion

- Use an attention-getting opening sentence.
- State your offer clearly.
- Be specific.
- Indicate the benefits of the service.
- In the closing statement, motivate customers to take action.

Model Letter of Reply to a Complaint

Date

MS EVA JOHNSON
5872 JACKSON PLAZA
SPOKANE WA 99200-3274

Dear Ms. Johnson:

Buffer statement

Logical statement

Positive tone

Our firm is committed to offer its customers quality merchandise at competitive prices. Therefore, it is important that our inventory control be maintained. To achieve this, we have a policy to give cash refunds on merchandise that is returned within 30 days of purchase.

Stress firm's policy

The garment you returned was purchased three months ago; therefore, our sales associate was only able to offer you a store credit and not a cash refund. To avoid any misunderstanding, the refund policy is posted at each register. We trust that you understand the company's position.

Goodwill gesture

End positively

We invite you to continue shopping at our store and have enclosed a 10 percent discount coupon toward your next purchase.

Sincerely yours,

Chris Adamo
ri

ACTIVITY F WRITING POWER

CASE STUDY 1: REWRITE OF POORLY WRITTEN LETTER OF REPLY TO COMPLAINT

Richard Santana attempted to return merchandise that he purchased more than 30 days ago. The salesperson displayed a negative attitude and did not explain the store's policy on returned merchandise. A poorly written letter was sent to Mr. Santana by the manager. After reading it, respond to the questions, then rewrite the letter. Send the letter to Mr. Richard Santana, 15 Lotus Street, Nashville, TN 37200-8426.

Case:

Dear Mr. Santana:

I received your letter describing the poor attitude that was shown to you by our salespeople when you returned some merchandise. This was wrong behavior.

Our store's policy is to give store credit only if the merchandise was not returned before 30 days after date of purchase.

We know you will understand.

Sincerely yours,

Timothy Jenkins

Manager

ri

Enclosure

Questions

1. What type of letter appears on the previous page?

 ____

2. Is the first sentence a good buffer statement? Why?

 ____

3. Has an apology been offered in the letter?

 ____

4. Does the letter have a sympathetic tone?

 ____

5. Is there a statement of goodwill to offset the customer's poor experience?

 ____

6. What type of closing statement would get results?

 ____

 ____

 ____

CASE STUDY 2: REPLY TO CUSTOMER COMPLAINT

Read the case below, then answer the questions that follow. Using your answers, respond to Mrs. Jessica Silvers, One Poplar Drive, Springfield, IL 62705-2265.

Case: Your customer, Mrs. Jessica Silvers, has contacted you regarding the damaged office desk that was delivered to her. As service manager, inform her that the damage occurred in transit from the warehouse and offer to send her a replacement of the desk. If Mrs. Silvers accepts this offer, she will receive the replacement within three weeks. At the time of delivery, the damaged desk will be picked up. However, if Mrs. Silvers is not satisfied with this arrangement, you can make an alternative offer of a desk of comparable quality and style. Ask Mrs. Silvers to contact your office with her decision. The telephone number is (800) 239-5000.

Organize Your Ideas

1. What is the most important point that you need to tell Mrs. Silvers?

___

___

___

2. What would you want to include in the opening sentence?

___

___

___

3. What specific details would you include in the body of the letter?

___

___

___

4. What would you say in the closing statement to indicate your desire to serve Mrs. Silvers?

___

___

___

CASE STUDY 3: PROMOTIONAL LETTER

Read the case below and compose the letter.

Case: Send a form letter to prospective customers inviting them to apply for your store's credit card, Buying Unlimited. Invite them to complete the enclosed application form to enroll in your membership program. Inform them that new subscribers will receive a $50 merchandise check that is applicable to any purchase made. An additional advantage for credit card holders is a 2 percent cash bonus for the total amount of their annual purchases.

CASE STUDY 4: REPLY TO AN INQUIRY

Read Case Study 4; develop an outline to guide you in writing, then compose the letter.

Case: Anthony Torres was given a rain check for stackable wall units that were unavailable. Mr. Torres is concerned that his rain check will expire and will not be honored and that he will not receive the merchandise at the sale price. As manager of the store, reply to Mr. Torres's letter, reassuring him that the merchandise will be sold to him at the sale price when it is received in the store. You may want to mention that the delay in filling the orders has been caused by the slow shipment from the manufacturer. Address the letter to Mr. Anthony Torres, 1301 Palma Drive, Houston, TX 77000-5487.

CASE STUDY 5: INFORMATION LETTER

Read Case Study 5; then compose the letter.

Case: This letter is in response to customers of Bailey's Department Store who are concerned with the release of personal information. They want to be sure that their files are not distributed to outsiders.

Jan Williams, president, would like you, her assistant, to prepare a letter to be sent to all Bailey's customers to assure them that no personal information is sold or distributed to anyone outside the firm. Bailey's has taken steps physically and electronically to maintain the privacy of all personal information of their customers.

Be sure to remind your customers that Bailey's Department Store provides quality namebrand merchandise at reasonable prices. Thank them for their loyalty.

CASE STUDY 6: E-MAIL MEMORANDUM

Read Case Study 6; then complete the e-mail memo.

Case: David Roche, executive vice president of Bailey's Department Store, is planning to hold a storewide inventory sale in July. He needs to meet with department managers to discuss the strategy for this sale. Some of the questions to be addressed are: What merchandise is to be put on sale? What type of discount should be offered? What type of promotional campaign will be used? What type of advertising—newspapers, home flyer, and/or TV—will be adopted? How long should the sale run?

Mr. Roche expects each one of the department managers to have a proposed plan based on these questions. Their input is important for the success of the sale. The meeting will be held on May 1 in the conference room.

Compose an e-mail memorandum to be sent to department managers. The subject of this e-mail is Planning Meeting for July Inventory Sale.

ACTIVITY G REPORT WRITING POWER: PERIODIC MEMO REPORTS

Periodic Reports

Periodic reports, one of the most common types of reports in business, often take the form of **memo reports.** They are prepared at regular intervals, such as monthly, quarterly, or annually.

The sales report is an example of a periodic report. To present data effectively in a report, tables and charts are often included. Sideheadings also contribute to the orderly flow of information.

Writing Pointers—Periodic Memo Reports

- Sales and financial reports are commonly written in memo form.
- Analysis of business operations can be reported internally in memo format.
- Periodic reports are prepared at specified times (i.e., quarterly, monthly).
- Charts and tables help the reader visually understand data.
- Statistical data is presented in table form.
- Informational statements should interpret data for the reader.
- Bulleted statements are used where clarification is essential for understanding.
- Reports are concise and clear.

Model Periodic Memo Report

TO: Michael Carpenter, Vice President

FROM: Rose Dimitri, Sales Manager

DATE: Current Date

Purpose identified

SUBJECT: Fourth-Quarter Sales

Year-End Results

Data presented

Our firm continues to show modest sales growth for the current fiscal year. From the beginning of the year, the sales increased by 33 percent. In the fourth quarter, sales increased 10 percent over the previous quarter. The following table shows a comparison of the sales figures for the past four quarters of the year. It is apparent that there is a pattern of growth.

Table format for clarity of statistical information

Quarter	$Sales	Percent of Increase/Decrease From Previous Quarter
Fourth	$418,000	+10.0%
Third	380,000	+22.5%
Second	310,000	−1.6%
First	315,000	+3.0% (from previous year)

Heading orients reader to content

Analysis

The fourth quarter increase in sales reflects an upward trend that began during the third quarter. A number of factors contributed to this turnaround:

Bulleted statements used for clarity

- Increase in seasonal buying of clothing
- Early start of cold weather during the fall months

- Increase in promotional activities
- Consumer interest in our classic, casual apparel
- Five percent decrease in general and administrative expenditures

The lower sales figure during the second quarter was unexpected. One of the reasons was the poor sales throughout the retailing industry due to a weak economy. With colder weather during the fall months, the demand for heavy-weight outerwear increased. Our new jeans collection in a variety of fabrics was introduced in September, and it has been very popular among our customers.

Interpretation of data clearly stated

During the past three months, we have refocused our advertising strategies to portray a brand image that identifies with high-quality merchandise. Newspaper advertising and mail-order flyers have been redesigned to illustrate to shoppers that we offer a versatile, yet classic style of clothing.

In addition, implementation of cost-saving procedures in shipping and distribution of merchandise has realized a 5 percent decrease in expenditures. With the use of a more sophisticated computerized system, our inventories have decreased and our delivery schedules are being met.

Conclusions

Sales projections in the retailing industry are very promising for next year; therefore, with our present plans, we should be able to continue to show an increase in profitability. We have moved forward as a company in building customer loyalty.

CASE STUDY 1: MEMO REPORT

After reading the case below, respond to the questions that follow to help you organize your ideas, then compose the periodic memo report. Address the memo to Mark Cohen, Vice President, from Eugene Dinsmore of the Assessment Team.

Case: Bayport, Inc., is a well-known catalog retailer. It offers an extensive line of women's clothing, footwear, and accessories. The clothing appeals both to the working woman and to the individual with a casual lifestyle. The semiannual catalog is the major medium used to advertise Bayport's merchandise, and seasonal flyers focus on specific items popular at various times of the year.

A review of the firm's operations was made to determine the catalog business and sales trends for the past three years. This resulted in the implementation of the following changes. First, a refocus of the firm's marketing strategies included the redesigning of its semiannual catalog with greater use of graphics and color to highlight merchandise. Second, a line of home merchandise directed at the needs of the working woman was introduced. Among the items were bath products, linens, and time-saving kitchen utensils. Third, a concerted effort was made to provide good customer service, including seven-day mail-order service.

The catalog circulation, which increased 22 percent over the past three years, was 50,000 copies for the current year; 45,000 and 41,000 for each of the previous years. The largest increase occurred this year when the changes were implemented. The number of active catalog customers who made at least two purchases a year doubled over the past three years to 20,000 customers. For each of the previous years, the number of customers was 12,000 and 10,000 correct respectfully.

The net sales during the three-year period also showed gains, especially during the current year with $1.6 million; net sales were $750,000 and $600,000 in each of the previous years.

As a member of the assessment team involved in the review, prepare a report of the firm's catalog business over the past three years and an analysis of the sales trend for the same period. Use a table to present the catalog and sales data for the past three years.

Organize Your Ideas

Note: This is your guide to writing your memo report. These questions follow in sequential order to help you develop the process of writing this memo.

1. What is the purpose of this memo report?

____

2. What is the subject line?

____

3. What is the best way to show a three-year comparison of Bayport's catalog business?

____

4. What are the three categories you would use in column 1 of the table?

   1. ____
   2. ____
   3. ____

5. What are the four column headings in the table?

   1. ____________________ 3. ____________________

   2. ____________________ 4. ____________________

6. How would you clarify important information that relates to a specific item in the table?

____

7. After completing the table, what is the sideheading of the next section?

____

8. What specific areas would you discuss in this section?

____

9. What three specific areas of change affected sales?

   1. ____
   2. ____
   3. ____

10. How would you incorporate in your report the listing of the three areas of change?

____

CASE STUDY 2: MEMO REPORT

The Corner Store, a specialty home furnishing retail chain with 150 stores nationwide, has undertaken a five-year expansion program. Prepare a memo report to the Executive Board reviewing the firm's progress to date. For clarity, the use of tables is recommended.

Case: Over the past year, the Corner Store's total sales for the geographic areas of the country were as follows: West Coast, 35 stores, $6.3 million; East Coast, 20 stores, $4.5 million; South, 5 stores, $675,000; Midwest, 10 stores, $1.2 million. The total increase in sales from the previous year was 3.5 percent, a small increase when compared to projections.

As part of the expansion plan, a study has been completed that evaluated the following areas: (1) analysis of sales, (2) review of product offerings, and (3) strategies for future growth.

The store brand name of towels and linens is very popular in all the stores. Customers also like the coordinated sets of sheets, bedspreads, and window furnishings. The bathroom products that sell in stores on the West and East Coasts are the best sellers. In the Midwest, sheets and pillows are excellent sellers; in the South, accessory items for the bathroom are very popular. The weakest sales in all stores are curtains and drapes. Customers have requested that the stores carry more accessory and gift items for the home.

Another area of expansion that needs to be considered is online shopping. Many of the competitors have web pages, and marketing studies show that retailers need to offer Internet services to survive.

Store displays lack a trendy image. A design consultant could make recommendations for better store layouts and displays of merchandise.

Over the next two years, the focus should be on improving sales in the existing stores rather than on rapid expansion of new stores. The opening of new stores should be limited to three in the West, five in the East, three in the South, and two in the Midwest.

The projected sales for the company should be a 10 percent increase for each of the next two years. An added line of house gifts and accessories should be considered; the slow sellers, curtains and drapes, should be discontinued unless these items are part of coordinated sets.

After the two-year period, stores not meeting the sales goals should be closed. Within five years, the Corner Store should have increased sales.

Organize Your Ideas

Note: Here is a guide to writing your memo report. The questions below follow in sequential order to help you develop the process of writing this memo. Remember to include a sideheading for each of the sections.

1. What is the purpose of this memo report?

   ____

2. What is the subject line?

   ____

3. What is the best way to show the breakdown of sales for each geographic area?

   ____

4. What column headings would you use in a table?

   ____

5. How should the best-selling items be listed?

   ____

6. List the recommendations you would make to improve the firm's sales.

   ____

   ____

   ____

   ____

   ____

   ____

7. What is the most effective way to list these recommendations?

   ____

ASSESSMENT: CHAPTER 7

English Power

Directions: *Insert commas where required.*

1. The business seminar that always presents timely topics will focus on the franchising industry.
2. Many shoppers who have limited time prefer to make purchases through a mail-order catalog.
3. Most department stores have sales during the Memorial Day weekend a popular time for shoppers.
4. We invite you our preferred customer to shop at any of our stores for extraordinary savings.
5. The knowledgeable friendly salesperson assisted the customer in choosing the gold bracelet.
6. On April 1 20xx the chain of video stores began its liquidation sale with a 50 percent discount on all items.
7. Employees in retail sales need a wide range of skills that includes merchandise knowledge people management and computer proficiency.
8. "Brushing Up on Your Business Skills" our one-day seminar will be given in Buffalo Pittsburgh and Toledo during the month of May.
9. Dr. Sandra Perez who has just returned from South America will address the Business Coalition Association on the topic of global selling.
10. Your mail order which was received today is backlogged for at least two weeks.

Editing Power

Directions: *Edit, format, and proofread the following unarranged letter for errors; key it correctly.*

Mr. Gregory Recio, 22 La Cholla Boulevard, Tucson, AZ 85705-3312

Dear Mr. Recio: We believe that quality performance and value makes our product outstanding. However we need to know from you our customer how your rate our product. A new Advisory Board has been formed to study changes that need to be made in our line of automobiles to better meet the needs of all customers. We invite you to answer a short survey asking for your views about added or changed features in our line of premier automobiles. As an owner you are in the best position to share you experience with us. We would also like to send you a free informative brochure titled Helpful guide when Purchasing a Car. This brochure contains worth while information which you can use when shopping for your next vehicle. The better informed you are as a consumer, the wiser your decisions will be. In the meantime we invite you to visit your dealer to test drive our latest models. Sincerely, Amy Morales.

Writing Power

Directions: *After reading the case below, draft a letter.*

As store manager of Allied Office Supplies, you are to notify your customer, Mr. Raheem Harris, of Business Products, Inc., 70 Cross Plaza, Detroit, MI 48232-9800, that his back order of office supplies has arrived. The purchase order number is 1285G. The order will be shipped immediately by UPS and should arrive within one week. Remember to apologize for the inconvenience.

> "Leaders don't force people to follow—they invite them on a journey."
>
> —*Charles S. Lauer*

Check-In with the Semicolon

WHAT YOU WILL LEARN IN THIS CHAPTER

- To punctuate independent clauses not joined by a conjunction
- To punctuate independent clauses in which either clause has an internal comma
- To punctuate independent clauses in which the second clause is introduced by a conjunctive adverb
- To punctuate a series of items that contains internal commas
- To develop a business vocabulary in the **HOSPITALITY INDUSTRY**
- To proofread and edit correspondence and reports
- To compose a letter of inquiry
- To compose a letter of response to an inquiry
- To compose a letter of complaint
- To compose an interoffice memorandum
- To compose an internal business proposal
- To compose a promotional letter
- To write an e-mail memo

Corporate Snapshot

HILTON HOTEL CORPORATION is the most recognized name worldwide in the hospitality industry. It is known for its fine accommodations and quality service. The Hilton chain includes four-star hotels, mid-priced accommodations, extended-stay facilities, resorts, and convention properties. Names such as Waldorf-Astoria in New York City, Hampton Inns, and Homewood Suites are easily recognized by business and leisure travelers.

Hilton Hotels are committed to preserving the environment. Through its Outreach Project, the firm participates with the community in planting trees, cleaning beaches, and restoring natural resources. Within the company, the goal is to reduce, reuse, and recycle. Programs have been implemented to reduce water usage, conserve electricity, and recycle products. The hotels also encourage their guests to conserve and reuse items.

In all aspects of the business operations, diversity is a priority. The corporation has adopted diversity principles into all phases of its business, including employment, training, management, marketing, purchasing, and advertising. It is management's belief that diversity has strengthened the organization.

Source: *http://www.hiltonworldwide.com*

ACTIVITY A LANGUAGE ARTS RECALL

Semicolon Usage

In Chapter 5, you learned that a comma is used before a conjunction that joins two independent clauses.

Example: We have landscaped our property, ***and*** these illustrations will show you our new appearance.

Rule 8.1: When there is **no** conjunction between two independent clauses, use a semicolon.

Example: We have landscaped our property***;*** these illustrations will show you our new appearance.

Rule 8.2: When there is a comma in both or either one of the independent clauses in a compound sentence, place a semicolon before the conjunction.

Example: To introduce a new product, you need to determine the target market; and you have to plan your marketing strategies.

Rule 8.3: When items in a series contain internal commas, use a semicolon to separate items.

Example: Popular ski resorts are located in Killington, Vermont; Aspen, Colorado; and Hunter, New York.

Rule 8.4: When terms such as *for example, namely,* or *for instance* introduce a list of items, use a semicolon before the term and a comma after it.

Example: Hotel chains are expanding their consumer markets; ***namely,*** resorts, health spas, and convention facilities.

Rule 8.5: When two independent clauses are connected by a conjunctive adverb (connecting word), use a semicolon before the conjunctive adverb and a comma after it.

Examples of conjunctive adverbs: however, nevertheless, furthermore, otherwise, thus

Example: Your health club membership expires this month; ***however,*** renewals at the discount rate will be extended for two weeks.

Check for Understanding

Directions: Review the following sentences for differences in punctuation, then answer the questions that follow.

1. You will receive a 20 percent discount at any of our hotels, and there is no cancellation charge.

2. You will receive a 20 percent discount at any of our hotels; there is no cancellation charge.
3. You will receive a 20 percent discount at any of our hotels; furthermore, there is no cancellation charge.
4. If you will make your reservation 30 days in advance, you will receive a 20 percent discount at any of our hotels; and there is no cancellation charge.

Questions

1. Compare sentences 1 and 2. Why does the punctuation differ?
2. Why does sentence 3 have both a semicolon and a comma?
3. Why is there a semicolon instead of a comma before the conjunction "and" in sentence 4?

(Answers are given below.)

Answers to "Check for Understanding"

1. Sentence 1 has a conjunction that joins two independent clauses. In sentence 2, there is no conjunction; therefore, a semicolon is used.
2. The conjunctive adverb "furthermore" connects the two independent clauses.
3. There is a comma in the first independent clause.

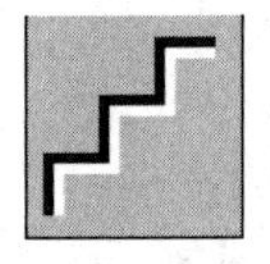

Building English Power: Commas and Semicolons

Directions: Insert commas and semicolons where required, then key correctly.

1. For hotel reservations you can call your travel agent or you can dial our toll-free number. (6.5, 8.2)
2. Hotels in major cities offer complimentary features in their weekend packages for example continental breakfast health club and free parking. (8.4, 7.1)
3. Popular vacation resort areas include San Juan Puerto Rico Cancun Mexico and Hilton Head South Carolina. (8.3)
4. A great way to visit new areas is to stay at bed-and-breakfast inns they are informal and friendly. (8.1)
5. The conference room has been reserved for our June meeting however plans for the luncheon are incomplete. (8.5)

6. The convention activities will be held in Scottsdale Arizona in addition local tours will be offered. (7.9, 8.5)
7. For your information a layout of the Conference Center is attached and a brochure about our facilities is enclosed. (6.5, 8.2)
8. Registration for the conference will take place in the hotel lobby and the exhibit area which is rather large will be held on the mezzanine floor. (8.2, 7.2)
9. We unfortunately cannot accommodate your group during the week of May 15 but our facilities are available the following week. (6.5, 8.2)
10. The members of the Executive Board after visiting several hotels met to discuss future convention sites but they were unable to reach a decision. (7.2, 8.2)
11. Reservations must be guaranteed with a credit card otherwise rooms are held only until 4 p.m. (8.5)
12. Our hotel chain combines quality accommodations fine service and good value moreover these hotels are highly rated by travel agents. (7.1, 8.5)
13. A block of 20 rooms has been held for your company conference you are reminded of our 48-hour cancellation policy. (8.1)
14. We offer a special weekend rate however reservations must be made two weeks in advance. (8.5)
15. Our hotels are located at the following airports: London England Milan Italy and Tokyo Japan. (8.3, 7.9)

ACTIVITY B BUILDING WORD POWER

The words below are commonly used in the field of hotel management. From this list, select the word or phrase that correctly completes each statement. Refer to the Personal Dictionary in the Appendix for the meaning of each word.

amenities	directories
complimentary	gourmet
concierge	implemented
consolidation	partition
continental breakfast	projections

__________ 1. A business trend in recent years has been the ________ of companies, which means that firms join together to eliminate duplicate departments and services.

__________ 2. Hotels offer guests many ________ so that they receive extra services such as free breakfasts and health club privileges.

__________ 3. ________ dining offers customers specialty foods often prepared with delicate sauces and seasonings.

__________ 4. Studies on future trends often present ________ based on the research findings.

__________ 5. New security measures were ________ after the accident in the lobby.

__________ 6. In many hotels, a(n) ________ copy of the local newspaper is delivered to guests.

__________ 7. The large room was divided by a(n) ________ to make two smaller rooms.

__________ 8. At airports, the traveler can find posted lists of hotels with their locations and rates on billboard-type ________.

__________ 9. Some hotels offer guests a(n) ________, which usually includes coffee and a roll.

__________ 10. The ________ assists guests in booking shows and sports events.

On the Alert

- compound adjectives—A one-thought modifier that describes a noun or pronoun is hyphenated.

  Example: check-in system, family-oriented clients, 24-hour access

- numbers—Numbers from one to ten are written as words.

  Example: ten meeting rooms

- "well" words—Use a hyphen to join the word *well* when used with another word to form an adjective before a noun; otherwise, it is not hyphenated.

  Examples:

  Our ***well-furnished*** rooms are popular. (hyphen)

  Our guest rooms are ***well furnished.*** (no hyphen)

ACTIVITY C INDUSTRY PERSPECTIVES: HOSPITALITY TODAY

Directions: Using appropriate revision marks, proofread and key the report below. The title of this report is "The Focus of Today's Hospitality Industry."

Rule Reference

Hospitality Industry in Today's Marketplace

Today's hospitality industry is complex and diverse. It is affected by social cultural political and economic changes. With a global economy conditions in one part of the world can have positive or adverse affects throughout the industry. In such a changing climate, the companies need to be prepared for any swing in business.

7.1
6.3

As a service business the hospitality industry's emphasis is placed on the individual. Employees are in direct daily contacts with clients from multinational cultures. It is critical for the workforce to be aware of other cultures and the necessity to be respectful of differences among people. Hotel chains therefore have made a commitment to promote diversity both within the organization and in the marketplace.

6.3
6.5

The hotels in this highly competitive field faces well informed, consumer-oriented buyers who demand high quality at good value. One of the most important assets for this consumer-directed business is goodwill. The brand name must reflect a company image of consumer satisfaction and excellent service.

3.1

Trends in the Hotel Business

Rule Referenc	
	Trends in the hotel business show an increase in
	leisure- and family oriented clients. To meet the
	demands of this class of clientele, hotels offer free
	breakfast, free local telephone calls, and indoor pools
	in addition room suites that accommodate families are
8.5	more readily available. Very often the hotel is in the
8.1	proximity of restaurant chains or shopping malls, this
	convenience can be a drawing card for the hotel.
8.4	Vacation packages continue to be popular namely resorts,
	health spas, and tours.
6.3	To continue to attract the business traveler hotels
	are adding business centers with 24-hour excess. These
	centers offer free usage of fax and copying equipment.
	The international market is a new source of business
6.1	for hotels. As the global market expands they must
	redirect their marketing strategies to these types of
2.3	clients. To maintain goodwill, staff need retraining in
	knowing the customs and preferences of foreign visitors.

Did you find these errors?
11 commas
3 semicolons
2 subject–verb agreement
2 word usage
2 hyphenated words

ACTIVITY D EDITING POWER

Task A: Edit, format, and proofread the following unarranged letter. Use proofreaders' symbols to mark the errors, then key the letter correctly. Remember to insert paragraphs.

Address this letter to Mr. Stuart Hall, 105 Los Gatos Boulevard, Sacramento, CA 95813–3223. It is from Mr. Carlos Ramirez, Convention Manager. The subject of the letter is "Facilities for Annual Convention." (See model letter in Appendix B.)

Errors
- semicolons
- commas
- subject–verb agreement
- word usage
- hyphens
- spelling
- typographical

We have recieved your inquiry about holding an annual convention at Royale Americana Hotel. We are pleased that we will be able to accomodate your group. Concerning our facilities they are the finest in the area and our convention rates are competitive. We have ten meeting rooms that hold a maximum of 50 people each and have noise proof partitions that allow for expansion. For the conference rooms the latest sound equipment is available. Our guest rooms are spacious and well furnished. The three gourmet restaurants that we have on the premises cater to different tastes however light snacks are served at our Oak Bar. We also offer our guests recreation amenities namely an indoor pool health spa and free movies. A walkway connects our hotel to a major shopping mall with boutique shops and well known retail stores. A computerized check-in system is used and our guests are greeted in the lobby with "smart-card" keys and a conference packet. A brochure price list and layout of our facilities is enclosed for your review. I am sure you will find Royal Americana Hotel the ideal sight for your convention.

Task B: Edit, format, and proofread the following unarranged rough-draft memorandum for all errors; key it correctly.

Send this memorandum to all hotel managers from J. Edward Holmes, General Manager. The subject is a "Review of Fire and Emergency Procedures." Use current date.

Recent events in the passed few weeks force us to

fire and emergency

review the ^ procedures in each of our hotel. I request

that you meet with your staff to discuss these ~~fire and~~

~~emergency~~ procedures currently being used. All

and tested

emergency equipment should be inspected ^ for proper

functioning. Since we are committed to having state-of-

the-art systems in all of our hotels you need to

determine whether the equiptment should be updated or

replaced. If you have any recomendations please outline

them for our review and include pertinent data to

support these suggestions. At a staff meeting discuss

sensitivity and

with the employees the need for ^ awareness of situations

that may occur in the day-to-day operations at your

hotel. You may find that some in-house training should

be given to the employees so that they will be well

prepared in case of an emergency. During these training

stet sessions clearly define the roll of key individuals in

your organization who are responsible for implementing

emergency procedures. A written report on the

suggestions and procedures should be submitted to my

office by teh end of the month.

ACTIVITY E WRITING PRACTICUM

Misplaced Modifiers

Modifiers are words that show a relationship with other words in the sentence. To be certain the meaning is clear, modifiers must be placed close to the word(s) they modify. **Misplaced modifiers** make the meaning of a sentence unclear and can even change the meaning of a sentence because the modifiers are too far removed from word(s) to which they relate.

Examples

Misplaced: Many of the larger hotel chains give a packet to their guests ***containing a room key and local information.***

Meaning? Does the writer mean that guests who have a room key and local information get a packet?

Correct: Many of the larger hotel chains give to their guests a packet ***containing a room key and local information.***

Misplaced: ***To reward employees for excellent service,*** an incentive program has been created by management.

Meaning? Is the incentive program rewarding the employees?

Correct: An incentive program has been created by management ***to reward employees for excellent service.***

EXERCISE 1 MISPLACED MODIFIERS

Correct the following sentences for misplaced modifiers. Change or delete inappropriate words. Use the blank lines provided.

1. The hotel visitors saw the golf course riding the jitney.

 ____

 ____

2. Mr. Smith left his briefcase in the hotel room containing his wallet, his passport, and his important papers.

 ____

 ____

3. Ms. Garcia met a relative she hadn't seen since childhood in the hotel lobby.

 ____

 ____

4. The new hotel was built in 1996 by Gutierrez Construction Co. and managed by Mr. Bright at a cost of $2 million.

 ____

 ____

5. The office clerk made a reservation for a room with double beds for Mr. and Mrs. Robert Jiminez with a private bath.

 ____

 ____

6. The Belle Convention Center has excellent facilities and good restaurants for the convenience of its guests with shuttle service to and from the airport.

 ____

 ____

7. Telephone reservations for prospective hotel guests are accepted on a space-available basis.

 ____

 ____

8. The dress code for dinner is formal for men with jackets.

____

____

9. Picture the landscaped beachfront vacation spot at the Waterfront Inn with its rocky shoreline.

____

____

10. We have deluxe hotels at two airports used by business travelers; namely, Chicago O'Hare and Los Angeles International.

____

____

11. The small hotel within walking distance of the shopping district is a charming place.

____

____

12. Spas for stressed individuals are ideal.

____

____

13. The concierge recommended a short distance from the hotel a restaurant to the guests.

____

____

14. The beach resort gave a complimentary night to its guests.

____

____

15. The resort offers its guests, nestled in the mountains, a relaxed vacation.

____

____

Dangling Modifiers

A **dangling modifier** is a modifying word or phrase that does not modify a word in the sentence.

Examples

Incorrect: ***While dancing in the nightclub,*** the fire alarm went off. (The italicized phrase does not modify a word.)

Correct: ***While dancing in the nightclub,*** we heard the fire alarm go off.

OR

While we were dancing in the nightclub, the fire alarm went off.
(The phrase modifies subject "we.")

Incorrect: ***Reading on the terrace,*** the noise from the highway disturbed me.

Correct: ***While I was reading on the terrace,*** the noise from the highway disturbed me.
(Changed the phrase to an adverbial clause.)

EXERCISE 2 DANGLING MODIFIERS

Rewrite the following sentences by correcting any "dangling" words or phrases.

1. I saw riding the hot-air balloon a spectacular view of the countryside.

   ___

2. While traveling in the Southwest, her family was touring Europe.

   ___

3. Hidden in the basement for the past 40 years, the owner of the sculpture decided to sell it.

   ___

   ___

4. While a guest at the hotel, the president arrived in town.

____

5. Visiting the seaside village many times, plans were drawn for a summer cottage.

____

6. Raining for three days, the visitors left the national park.

____

7. While getting ready for the staff meeting, the report was completed.

____

8. After a 10-hour flight, fatigue set in.

____

9. With its breathtaking scenery, the people of New Zealand welcome visitors.

____

10. Preferring the beaches to the mountains, the Carribean was our choice.

____

EXERCISE 3 MODIFIER ERRORS

Rewrite the following paragraph that was taken from the brochure of a new hotel management school. Clear up any confusion created by misplaced modifiers. You may have to add words for clarity. For ease of checking revisions, each sentence has been numbered.

(1) Anyone who has studied at the Acme School of Hotel Management for a matter of minutes can learn to make quick decisions in the field of hotel management. (2) Located in a thriving and bustling downtown metropolitan area, our students benefit from internships at nearby hotels. (3) Well-trained and professional, our workshops and seminars are led by leaders in the hotel industry. (4) Our brochure, which includes itemized costs and course descriptions, on request will be free of charge.

EXERCISE 4 MODIFIER ERRORS

The following paragraph is taken from a travel diary. Rewrite it, clearing up confusion created by misplaced modifiers. Insert extra words where needed for clarity. For ease of checking revisions, each sentence has been numbered.

(1) Arriving at our hotel, members of a Bird Watchers' Convention had been checked into our room. (2) Apologizing, we were led to another room by the desk clerk on the first floor. (3) The dining room was crowded with people eating with binoculars and discussing the local migrating population. (4) There were other guests among the bird watchers who were hard to spot. (5) Rising early to spot the birds, we were awakened by the devotees each morning. (6) Eventually giving in, the bird-watching excursions intrigued us, and we decided to join in on the fun.

Letters of Inquiry

Letters of inquiry request information. The different types of inquiries are

- request for appointments and reservations
- request for purchases
- general request

All types of inquiry letters should give the reader the necessary information for a quick and complete response.

Writing Pointers

Letter of Inquiry

- Request specific information (for example, facilities).
- Give all dates, location, and time when requesting appointments.
- State special services that you may need.
- Use questions to obtain information

Letter of Response to an Inquiry

- Start with an upbeat tone.
- Consider reader a prospective customer.
- Give exact information requested.
- Highlight positive points.
- Express appreciation for the inquiry.
- Show potential benefits.
- Offer to provide further assistance.

Letter Format

- Use number or bullet format to focus on important facts or to list important points. (See Appendix B.)

ACTIVITY F WRITING POWER: COMPOSING CORRESPONDENCE

CASE STUDY 1: RESPONSE TO A LETTER OF INQUIRY

After reading the case below, respond to the questions that follow, which will help you organize your ideas before beginning to compose the letter. Then draft the letter to Mr. and Mrs. Donald Russell, One Poplar Drive, Salt Lake City, UT 84119–0050. (Organize your ideas on p. 208)

Case: As a reservation representative at the Cameo Bay Club, a resort hotel, send a response to Mr. and Mrs. Donald Russell on their inquiry about your resort. Cameo Bay Club is a luxurious resort with many amenities. In your letter, you may want to mention some of the benefits Mr. and Mrs. Russell would enjoy as guests at the Cameo Bay Club. Remember to enclose a brochure and price list.

Model Letter of Inquiry

Date

MS CHERYL CASEY
VICE PRESIDENT SALES
SUN HOTELS
1401 BISCAYNE BOULEVARD
MIAMI FL 33100-8312

Dear Ms. Casey:

Positive opening related to hotel services

Many of our employees choose your hotel chain while traveling on business. They find that at each of your hotels, the service is outstanding and the staff always helpful. In addition, you offer many amenities that our people need when away from the home office.

Specific rationale and facts given for a response

Request for specific information

It is important for our personnel to be able to work effectively when they are conducting business out of town. Therefore, we are considering using your hotel chain whenever our employees are on business. Last year, our staff booked a total of five hundred nights with your chain. We are interested in working out an agreement to receive discounts on room reservations for our personnel.

Request for a reply

I would be happy to meet with you to discuss such an arrangement. Please call me at 1-407-932-6654.

Sincerely yours,

Marcia Keane
General Manager

ri

Model Letter of Response to an Inquiry

Date

MR AND MRS BERT LATHAM
2205 SUNSET DRIVE
AUSTIN TX 78700-3521

Dear Mr. and Mrs. Latham:

Welcome to the Bay Inn!

Upbeat tone

Our beautiful resort overlooking the Peconic River is a wonderful vacation spot. The highly rated Bay Inn has once again received the Traveler's Magazine Award for Outstanding Service.

Highlights important features

To help you plan your vacation, we have enclosed our resort brochure describing our amenities and area attractions. Within minutes from the Bay Inn, there are gourmet restaurants, boutique shops, and recreational theme parks. Our concierge is always ready to answer questions and to assist you in planning local activities.

Cordial closing

Further assistance offered

We invite you to book with us and enjoy a memorable vacation. Feel free to call us at 800-237-4100 or visit our web site at www.bayinn.com.

Sincerely yours,

Ellen McCool
Manager

ri
Enclosure

Case Study 1 continued

Organize Your Ideas

1. What is the most important point you want to make to Mr. and Mrs. Russell?

____

____

2. What would you include in the opening sentence?

____

____

3. What are some of the benefits that the Cameo Bay Club offers?

____

____

____

4. What would you say in the closing statement to get Mr. and Mrs. Russell to book a reservation?

____

____

CASE STUDY 2: LETTER OF INQUIRY

Read the case below, then answer the questions that follow. Use bullet format to highlight important points.

Case: As the administrative assistant to Ms. Betty Ryan, you are writing a letter inquiring about facilities for a three-day meeting. We plan to hold this meeting on June 5 to 7 at the Greenpoint Hotel. Find out what the rates are for a conference room. We will have approximately ten people. We also need five double guest rooms that are nonsmoking. What will these rates be? Please check the cost of a coffee cart for both the morning and afternoon. You also need to find out if an overhead projector and screen are available. Send the letter to Sales Department, Greenpoint Hotel, 22 River Road, Seattle, WA 98109–0900.

Organize Your Ideas

1. What is the most important point you want to make to the sales representative at the Greenpoint Hotel?

____

____

2. What would you include in the opening sentence?

____

____

3. What specific details would you include in the body of the letter?

____

____

____

4. What would you say in the closing statement to get an immediate response?

____

____

CASE STUDY 3: COMPLAINT LETTER

Read the case below, then write a letter of complaint to Mr. Roy Foley, general manager at the hotel, 1125 Magnolia Drive, New Orleans LA 70100–2435.

Case: You were a guest at the Traveler's Hotel in New Orleans for three nights. During your stay, the air-conditioning unit was not functioning properly, which made the room uncomfortably hot. When you complained to the main desk, you were assured that it would be taken care of immediately. After repeated complaints, no service person came until the third day of your stay. The excuse was that the hotel was short of service personnel. The main desk offered no solution.

CASE STUDY 4: MEMORANDUM TO STAFF

Read the case below, then compose a memorandum to the hotel staff informing them of the holiday project to collect toys for needy children in the community.

Case: As the administrative assistant in the Human Resources Department, draft a memorandum to the hotel staff announcing a holiday project to collect toys for needy children in the community. The drive will run for three weeks in December, during which time the toys will be collected and displayed in the main lobby of the hotel.

The staff is being asked to adopt a slogan for the drive; therefore, try to be creative when appealing to your coworkers to contribute a toy for a worthy cause.

Send the memo from your manager, Gregg Simms.

(See the memorandum format in Appendix B.)

CASE STUDY 5: PROMOTIONAL LETTER

Read the case study; then compose the letter.

Case: Ken White, program director of Pathways to Diversity, is writing to Gerald Simpson, Vice President of Human Resources, about promoting a new employee training program for the development of positive attitudes toward cultural diversity. Mr. White points out the importance of offering such a program, especially in the hospitality industry where employees deal with diverse clientele. Employees must be good team members who respect each other and their customers. Pathways to Diversity includes in-service workshops and seminars. The workshops are case-oriented, discussing diverse customs of people in different parts of the world, sensitivity training to avoid conflict, and appreciation for other cultures.

A staff representative will gladly meet Mr. Simpson to make a formal presentation on the program. Compose a letter for Ken White. Mr. Simpson's address is Economy Lodge International, One Park Row, San Antonio, TX. Include subject line: Pathways to Diversity Program.

CASE STUDY 6: E-MAIL MEMORANDUM

Read the case study; then write an e-mail memo.

Case: Write an e-mail memo from Lori Reynolds to Gerald Simpson, Vice President of Human Resources (Gsimpson@Menzahotel.com). Lori Reynolds would like to meet with Gerald Simpson to discuss the possibility of starting a child day-care center on the premises of the Menza Hotel. A growing problem at the hotel is the increase of employee tardiness and absenteeism. The loss of man hours is costly to the firm. Many of the employees work nontraditional hours. As a result, it is difficult to find qualified help to take care of children. Lori's staff at Human Resources has researched the problem and has proposed a plan. She believes that a day-care center would be beneficial to the company by increasing productivity. There would also be an improvement in employee morale if parents can visit their children during lunch or dinner breaks.

The subject of this e-mail is Day-Care Center Proposal.

ACTIVITY G REPORT WRITING POWER: INTERNAL BUSINESS PROPOSAL

Business Proposals

A business proposal, which is commonly used in business and government, defines a need or problem and recommends solutions. The major purpose is to persuade readers to improve a product, to accept a service, to purchase equipment, to change procedures, or to solve a problem. A proposal can be initiated outside the organization or by internal personnel.

Writing Pointers

- Write an internal proposal in memo format.
- Know the purpose of the proposal.
- State the need or problem.
- Give details of need or problem.

- Present all necessary information to support your proposed plan.
- Provide accurate and complete data.
- Offer a persuasive solution to the need or problem.
- Discuss any staffing or budgetary needs.
- Use sideheadings to help readers focus on a particular point.
- Use tables or charts to present figures or percentages.
- Avoid first-person approach.
- Stress the benefits of the proposed plan.

Model Internal Proposal

TO: Members of the Town Council

FROM: Lois Branson, Council Representative

DATE: Current Date

Introduces topic

SUBJECT: PROPOSAL TO ESTABLISH A VISITORS' BUREAU

Purpose identified

With the major goal of the Town Council to form a partnership between business and local government for the purpose of enhancing the economic and social quality of life of Greenport, the opening of a Visitors' Bureau would be a first step in realizing such an undertaking.

BACKGROUND INFORMATION

Specifies details of problem

In recent years, many of the surrounding towns and cities have begun to revitalize their downtown areas and have been competing for new businesses and increased tourist trade. To remain competitive, Greenport must take steps to consider a master renewal program. To spearhead the program, we propose the establishment of a Visitors' Bureau. Such a bureau would work with the hospitality industry, local retailers, and service businesses in coordinating events, contacting media, and promoting Greenport as a "fun place to visit."

States accurate data

Presently, the hotel room occupancy in Greenport is only 65 percent. In the past year, two well-established retail businesses have closed. Over the past five years, there has been a decline in the number of business conferences and trade shows held in Greenport. This decline has had a negative effect both on the hospitality industry and on service businesses, such as restaurants, catering establishments, delivery services, auto rentals, and health facilities.

Stresses benefits of proposal

BENEFITS OF A VISITORS' BUREAU

With an increase in business and leisure trade, there would be job opportunities for local residents, added tax revenues, increased patronage at restaurants, higher hotel occupancy rates, and an overall positive outlook by our citizens. A Visitors' Bureau would have a direct impact on the economy of the city. As a government-sponsored agency, it would have the clout to meet with industry and encourage new businesses to consider Greenport. The bureau could cooperate with the hospitality industry to sponsor events, fairs, and tour packages. As a clearinghouse for tourist information, a Visitors' Bureau would be the ideal agency to advertise the benefits of working and visiting Greenport.

COST OF ESTABLISHING A VISITORS' BUREAU

Discusses costs of proposal

The cost of establishing a Visitors' Bureau would be minimal. For two years, the program could be initiated on an experimental basis. It is also recommended that local businesses and the hospitality industry help to defray some of the costs. The following are projected expenses:

Shows columnar format for clarity

Office space	Use of vacant space now available in Town Hall
Administrative expenses	One full-time administrator (salary: $25,000 per year) One part-time assistant (salary: $8 per hour)
Telephone/fax	$100 per month
Publication	In-house printing ($3,000 per year)

TASKS OF VISITORS' BUREAU

Discusses role of organization

Initially, the major goal of the administrator of the Bureau would be to make immediate contact with the hotels, motels, and local businesses in the city to form a partnership in promoting Greenport. After frank discussions of the needs of these establishments, some programs could be started. Among possible activities are the following:

Uses bulleted statements for clarity and conciseness

- Encourage specialty and craft shops to locate in downtown area.
- Sponsor a community fair.
- Offer hotel and restaurant packages to tourists.
- Advertise the amenities of Greenport.
- Publish brochures and maps on local attractions.
- Act as a liaison between government agencies and possible new industry.

RECOMMENDED ACTION

Requests action

We need to take action on this matter. Our Town Council should invite the business leaders of the community to an open meeting to receive reaction to the proposal. After this meeting, the Town Council would appoint a chairperson to work with a temporary planning committee to lay the groundwork for the formation of a Visitors' Bureau.

CASE STUDY 1: BUSINESS PROPOSAL

After reading the case below, respond to the questions that follow to help you organize your ideas, then compose the business proposal.

Case: You are the manager of the Omni Hotel and have observed over the past three years the increase in foreign guests who stay at the hotel. Further study showed that during the current year, 40 percent of the guests were foreign visitors. They represented 25 percent of business travelers and 15 percent of tourists. A further breakdown showed that 15 percent of all guests are Hispanic; 15 percent, European; and 10 percent, Japanese.

The hotel personnel at the reception desk, department supervisors, and service employees speak English only. There is a growing problem of miscommunications between hotel staff and foreign guests. This has caused misunderstanding, poor guest perceptions, and frustrations of staff. There have been unnecessary confrontations and disagreements between clients and staff.

Because of this problem, you are submitting to Ms. Lori Battle, Vice President of Operations, a proposal that addresses the situation and offers recommendations. First, all future employees in supervisory capacities and staff in front-desk operations should be bilingual or with multicultural backgrounds.

Second, an educational program should be initiated to offer language courses in Spanish and Japanese. All personnel should have the opportunity to take a course of their choice. As a pilot program, a 12-week course with classes meeting twice a week could be offered. This course should focus on everyday language that would pertain to hotel communications. In addition, a sensitivity training seminar that discusses the different social attitudes and behaviors of foreigners should be given to all hotel employees.

Third, pamphlets that explain hotel services, amenities, and local tourist information should be available in various languages.

The benefits of the proposed plan would be greater customer satisfaction, especially among foreign guests. Better communications and understanding is the best way to appeal to former guests to return to the hotel. The Omni Hotel would continue to maintain its fine reputation among travel agencies and individual clients as a customer-friendly place to stay when visiting the city. The nominal cost of initiating an education program would be returned in increased revenues.

To run the program, you would need a conference room at the hotel, a language teacher, and textbooks. The fee for a teacher would be approximately $50 per hour for classes of 10 individuals. The cost of the textbook is $25 per book. A speaker on multicultural customs and attitudes can be brought in for about $500 per two-hour session.

Let management know that if the Omni Hotel is to remain competitive, it must meet the needs of its clients. Recommend that the program start within six months.

Organize Your Ideas

To help you write an effective internal proposal, respond to the following questions below before composing it.

1. What would you use for the Subject Line in this proposal?

____

2. What facts should be included under the side heading "Background Information"?

___

3. What are the two major points that you would indicate under the benefits of the proposed program?

   1. ___
   2. ___

4. How would you present the key points of the plan?

___

5. What are the important points of the proposed plan?

___

6. How would you feature the cost of the program?

___

7. What information would you include in your answer to question 6?

___

8. What would you state in your final paragraph?

___

CASE STUDY 2: BUSINESS PROPOSAL

To remain competitive, the management of Hotel Gemini is exploring the possibility of offering weekend, three-day, or five-day promotional packages. Write a proposal in which you recommend a package deal. You may opt to offer only one of the three packages. In preparing the proposal, use your own experience, refer to newspaper ads for ideas, and be creative.

The following are some facts that may help you develop the proposal.

- Hotel Gemini is located in the center of the city near museums, sports facilities, theaters, exclusive shopping, and restaurants.

- Hotel bookings are low on weekends and during the months of January, February, and March.
- Hotel clients are 60 percent businesspeople and 40 percent tourists.
- The rate for a double room is $150 per night. A 15 percent discount is available.
- The hotel has a gym and health spa.
- Packages might include theater, concert, and sporting event reservations as well as restaurant and shopping discounts.

Organize Your Ideas

In this case study, the following questions are only guidelines. Feel free to add your own creative ideas to the promotional package you decide to offer.

1. What sections will you include in the proposal?

   ____

   ____

2. What is the need of the proposal?

   ____

   ____

3. Which package do you expect to use?

   ____

   ____

4. What will you include in your package deal?

   ____

   ____

5. How will the hotel benefit from this offer?

   ____

   ____

ASSESSMENT: CHAPTER 8

English Power

Directions: *Insert commas and semicolons where needed. Key sentences correctly.*

1. Popular hotels are usually booked to capacity during peak times therefore we recommend that you reserve in advance.
2. If you need assistance call our customer service and be sure to give your membership number.
3. We offer travel savings to hotels motels and resorts throughout the world for example Montreal Rio de Janeiro Puerto Rico and Hawaii.
4. On our seven-day package deals to the Caribbean you are entitled to an additional 25 percent discount there is a nominal fee for daily maid service.
5. For your convenience our 24-hour hotline informs you of the available resort accommodations therefore do not hesitate to call us at any time.
6. The travel agent who has lived in the Orient has recommended visiting Thailand and Singapore furthermore he is preparing a list of resort hotels.
7. Our hotel has been refurbished recently its rooms have been restored to their original decor.
8. Your itinerary has a detailed schedule of your trip namely hotels tours and meals.
9. Some of our hotels cater to specific customer needs for example Cancun resort complex Kansas City convention center and Orlando family entertainment.
10. Through our preferred membership program you can save 50 percent off hotel rates in addition you will be able to take advantage of our special vacation packages.

Editing Power

Directions: *Edit, format, and proofread the following unarranged letter. Supply missing parts and paragraphs; key it correctly. Address this letter to Mr. Frank Stabile, Longwood Corporation, 1469 East Main Boulevard, Kansas City, KS 66110-2215.*

I enjoyed meeting with you last Tuesday to finalize plans for your conference scheduled for March 17-20, 20xx. The facilities we agreed upon include the use of five meeting rooms equipt with a lectern and microphone. An overhead projector screen and flip chart will be available in each room. These rooms will accomodate 25 people at each session they will be free for your group from 8:30 a.m. to 5:30 p.m. each day. Our small dining room the Blue Room will be set up for diner for three evenings March 17 18 and 19. Mr. Jack Adams our banquet manager will contact you regarding the menu and room layout for each evening. As you know we have a fine reputation for our cuisine therefore we can offer creative ideas for afternoon breaks and a reception hour. I am enclosing a contract which specifies the conference facilities and arrangements. If the contract meets with your approval please return the three signed copies of the contract and enclose a deposit of $500. I look forward to working with you on final arrangements. Sincerely yours, Diane Somerset, Convention Manager.

Writing Power

Directions: *After reading the case below, draft a letter.*

As administrative assistant to Dr. Edythe Baxter, reserve a two-room suite for a meeting which can accommodate a group of six people. You will need this room for two nights, May 15 and 16. Indicate that you prefer a nonsmoking floor and that the room should have a view of the city. Also, make a reservation for dinner for six at the Ambassador, the hotel restaurant, on May 16 at 7 p.m. in the nonsmoking section. Request a confirmation of the suite and dinner reservations. Send the letter to the Reservations Department, Hotel Arrowood, 1005 Harbor Boulevard, Dallas, TX 75260–3921.

> "TEAMWORK: Simply stated, it is less me and more we."
>
> —*Heartquotes.net*

9

Marketing Possessives

WHAT YOU WILL LEARN IN THIS CHAPTER

- To apply possessives correctly
- To develop a business vocabulary in **MARKETING**
- To proofread and edit correspondence and reports
- To compose bad news correspondence
- To compose promotional letters
- To compose a memorandum
- To write an e-mail
- To compose an external business proposal

Corporate Snapshot

Ben & Jerry's. In May 1978, two childhood friends, Ben Cohen and Jerry Greenfield, opened an ice cream parlor business in a renovated gas station in Burlington, Vermont. This site was chosen in a college town without an ice cream shop. Ben & Jerry's soon became known not only for its unusual ice cream flavors but also as a community-oriented business committed to respect the environment.

The firm's mission statement focuses on the use of wholesome ingredients in its products and the improvement of the quality of life both locally and nationally. Business is only done with dairy farmers who do not give growth hormones to their cows. The company partners with nonprofit organizations to open ice cream shops that provide job training and employment opportunities to disadvantaged persons.

The company founders insist that "Business has a responsibility to give back to the community."

Source: *http://www.benjerry.com*

ACTIVITY A LANGUAGE ARTS RECALL

Possessive Case

Possessive words indicate ownership.

Example: the customer'***s*** needs (the needs of a customer)
the manager'***s*** report (the report of the manager; the report belonging to the manager)

A possessive word is a noun and may be a person, place, or thing. It may be singular (one person, place, or thing) or plural (two or more persons, places, or things).

Example: company'***s*** product (singular)
consumer***s***' lifestyles (plural)
Ramon'***s*** store (singular)

Singular and Plural Nouns

Rule 9.1: Form the possessive case of a regular singular noun that does not end in "s" by adding an apostrophe plus "s" (**'s**). This rule also applies to plural nouns that do not end in "s."

Examples: buyer'***s*** territory (s) men'***s*** clothing (pl)
child'***s*** toy (s) children'***s*** toys (pl)
manager'***s*** policy (s)

Rule 9.2: Form the possessive case of a singular noun that ends in "s" by adding only an apostrophe. However, if an additional syllable is pronounced after the "s," then add an apostrophe plus "s" ('s).

Examples: The Hasting***s***' plan
business'***s*** boss'***s*** Jones'***s***
(Note: A new syllable is formed in pronunciation; therefore, add '**s**.)

Rule 9.3: Form the possessive case of a plural noun that ends in "s" by adding an apostrophe after the "s."

Examples: customer***s***' buyer***s***' businesse***s***'

Compound Nouns

Rule 9.4: If the compound noun does not end in "s," add an apostrophe plus an "s" ('***s***). Form the possessive of a compound noun that ends in "s" by adding only an apostrophe.

Examples: vice president'***s*** office (office of the vice president)
vice president***s***' meeting (meeting of the vice presidents [pl.])
sister-in-law'***s*** gift (gift of the sister-in-law)
sister***s***-in-law'***s*** gifts (more than one sister-in-law; therefore, form the plural of the word first and then add '***s*** at the end of the compound noun)

Joint Ownership and Separate Ownership

Rule 9.5: To show joint ownership (ownership of the same thing by two or more persons), add an apostrophe plus an "s" to the last person's name. Ordinarily, if the name ends in "s," add only an apostrophe.

Examples: Carmen and Rosa'***s*** boutique shop (Both women own the shop.)

Frank and Joe'***s*** company (Both men own the same company.)

Rule 9.6: For individual ownerships, add an apostrophe plus an "s" after each of the names.

Example: Lisa'***s*** and Mario'***s*** pet shops are located in the South. (two pet shops, each separately owned)

Personal Pronouns

Rule 9.7: Do not use apostrophes in the possessive form of personal pronouns.

Examples: his, his; her, hers; your, yours; our, ours; their, theirs; its, its
Note: "It's" is a contraction meaning "it is."

Indefinite Pronouns

Rule 9.8: Some indefinite pronouns have regular possessive forms.

Examples: one's; other's, others'; anybody's

Check for Understanding

Directions: Indicate the singular and plural possessive form of the italicized words in the following phrases.

Example: *salesperson* territory (S) salesperson's territory
(P) salespersons' territories (or salespeople's)

1. *business* profit (S) ____________
(P) ____________
2. *man* clothing (S) ____________
(P) ____________
3. *class* discussion (S) ____________
(P) ____________
4. *manufacturer* list price (S) ____________
(P) ____________
5. *son-in-law* firm (S) ____________
(P) ____________

(Answers are given below.)

Answers to "Check for Understanding"

1. business's profit
businesses' profits
2. man's clothing
men's clothing
3. class's discussion
classes' discussions
4. manufacturer's list price
manufacturers' list prices
5. son-in-law's firm
sons-in-law's firms

Building English Power: Possessives

Directions: Insert an apostrophe or 's where necessary. Then key each sentence.

1. The firm decision to introduce video disks in its marketing plan will affect the advertising department. (9.1, 9.7)
2. Management met to discuss the use of telemarketing to increase the businesses sales. (9.2)
3. A multimedia approach provides opportunities to meet many readers and listeners needs. (9.6)
4. David and Emily proposal to study the advantages of direct mailing was given to the marketing research group. (9.5)
5. The applicant with five years experience in fashion display was offered the department manager position. (9.3, 9.1)
6. Franchises offer an individual the opportunity to be ones own boss. (9.1)
7. The Jones and Burns companies are merging to increase their productivity. (9.2, 9.6)
8. Salespersons attitudes often determine customers decisions to purchase merchandise. (9.3)
9. Competitors prices are influencing the sales policies of the firm and its subsidiaries. (9.3, 9.7)
10. The women and men departments are being reorganized to include designer clothing. (9.1, 9.6)
11. We bought three similar CDs for our brothers-in-law birthdays. (9.4)
12. The store is offering a 10 percent discount off the manufacturer suggested retail price. (9.1)
13. Trucks are a popular means of transporting goods because merchandise can go directly from the shipper place to the customers business. (9.1)
14. The library new electronic information center will offer access to business periodicals. (9.1)
15. Although Jared has the advantage of better locations for his shops than Eric, each of Jared and Eric repair shops has a good reputation for fast and satisfactory service. (9.6)

ACTIVITY B BUILDING WORD POWER

The words below are commonly used in the field of marketing. From this list, select the word or phrase that correctly completes each statement. Refer to the Personal Dictionary, if necessary.

commission	network marketing
entrepreneur	rebate
feasibility	standalone
franchise	subsidiary
franchisee	supplement
franchiser	telemarketing

_____________ 1. Independent owners acquire a(n) ________, which is a legal contract to own and operate a well-established business according to the regulations of the parent corporation.

_____________ 2. You, the ________, bought the licensing rights to own and operate the Pizza Delight store in San Diego.

_____________ 3. Salespeople in the automobile industry receive a small base salary plus ________ on their total sales.

_____________ 4. The term ________ often refers to computers that are not interconnected with large systems; for example, personal computers.

_____________ 5. Some large corporations have ________ companies that they control.

_____________ 6. Many people have a second job to ________ their income.

_____________ 7. ________ is a popular trend in which sales are made over the telephone.

_____________ 8. ________, also known as multilevel marketing, is a selling approach in which individuals sell products privately through contacts and recommendations.

_____________ 9. The owner of the local flower shop is an example of a(n) ________ in our economic system.

_____________ 10. The firm that sells franchise rights is referred to as the ________.

_____________ 11. The company is studying the ________ of introducing voice-activated disks as a workable approach in selling.

_____________ 12. To promote sales, the automobile manufacturer is offering a $500 ________ on new cars.

On the Alert

- assistance (noun)—the act of giving help or support

  Example: The firm gave financial ***assistance*** to its disabled employee.

- assistants (plural noun)—individuals who help or support

  Example: Both of the marketing ***assistants*** handled the sales records for the branch office.

- geographic locations—should be capitalized when referring to a specific region

  Example: West Coast; Northeast *BUT* travel north; southeast of Hartford

- ASAP—abbreviation for "as soon as possible," a common term used when shipping merchandise

ACTIVITY C INDUSTRY PERSPECTIVES: MARKETING AND PUBLIC RELATIONS

Directions: Using appropriate revision marks, proofread and key the report below. The title is MARKETING AND PUBLIC RELATIONS FOR THIS DECADE

Rule Reference	
7.1	Marketing is the tool used to inform the public of a
	product, service, organization or idea. Good marketing
	strategies translate into successful selling of
	merchandise or ideas. Marketing and public relations
	work hand-in-hand to maintain a positive image between
8.4	an organization and its public namely customers,
7.1	employees, shareholders, government and media.

Rule Reference

PUBLIC RELATIONS

The job of public relations is to convey credibility,
7.1 develop goodwill and enhance an organization's or
9.1, 9.7 individuals image. It's role is product publicity,
corporate publicity corporate publications, and creation
of a public image. The clients of a public relations
firm include corporations, charitable organizations,
7.1 government agencies, performers and politicians.
Organizations are very sensitive to the image that they
portray in the marketplace. Identifying with a brand
name, a logo, or a jingle is important when selling a
product or service. One does not realize that promoting
a singer or presenting a politician for an election is
similar to advertising a product. It is the
responsibility of the public relations firm to portray a
positive image.

MARKETING TOOLS

6.3 In this world of technology e-mail can be a powerful
marketing tool. It is especially popular with retailers
since it gives them an opportunity to build a good
6.5 customer relationship that translates into sales.
6.6 Furthermore e-mail is less costly compared to other
media advertising such as television. Some firms have

Rule Reference	
	initiated monthly e-mail newsletters to there customers.
	These newsletters not only give helpful information but
	promote products or services.
	An alternative to market merchandise or services is
	through the Internet. Most retailing firms have web
	sites for customers to order merchandise. This is a
9.8	convenient way to shop from the comforts of ones home
9.1	during an individuals free time. The travel and
	hospitality industry use the Internet to offer customers
	discount rates and package deals.
	Another marketing strategy to meet the demands of
6.3	busy shoppers is mail-order catalogs. Once again it is a
	convenient way to shop for any type of item without the
6.3	hassle of shopping at a mall. To be successful a mail-
	order firm must offer quality merchandise and good
	service. Mail-order firms rely heavily on well designed
	catalogs that display attractive photographs of the
6.3	products. In many cases the catalog resembles a travel
	book or fashion magazine. This marketing strategy
	attracts the potential customer to order merchandise.

Did you find these errors?
12 commas
1 semicolon
4 possessive
1 subject–verb agreement
1 hyphenated word
1 word usage

ACTIVITY D EDITING POWER

Task A: Edit, format, and proofread the following unarranged memorandum. Use proofreaders' symbols to mark the errors, then key the memorandum correctly. Remember to insert paragraphs.

Send this memorandum to the sales representatives from Eric Choy, Vice President. The subject is fax software.

Errors
- commas
- semicolon
- word usage
- possessives

```
Because of complaints from sales representatives we are
now considering alternative procedures to eliminate time
wasted while waiting to use a standalone fax machine.
One suggestion is the feasibility of using fax soft ware
to transmit documents directly from the desktop PCs.
Studies show that users of this type of software have
increased there productivity sales representatives can
save an hour of each workday. You no longer waste
valuable time that could be spent contacting a
perspective client. General computer Company will
present it's software at a meeting scheduled for Monday
June 19. Please clear your calendar to attend. After
this session we shall expect the sales forces reaction
before a decision is made.
```

Task B: Edit, format, and proofread the following rough-draft letter for all errors, then key it correctly.

Address this letter to Mr. Brian Price, Vice President at The Clothing Loft, Inc., 1150 Dogwood Lane, Tulsa, OK 74101–7562. This letter is from Muriel Ryan, General Manager. Use current date.

We will be happy to meet with you and your staff in exploring the possibility of offering our ~~consultants~~ expertise to your firm. Anderson & Andersons reputation as a consulting firm in the direct marketing field is well-known. The direct marketing business can be a successful profit-making venture. However the management skills in running this type of business is crucial. We ~~would~~ will help you define a plan ~~which could~~ that will work for you. After your staff identifies your consumers market we will design the most effective promotional program to built a consumer base. We also will explore with you ~~the~~ effective ways to sell through catalogs which have shown sales increases in the passed few years. Our staff representatives are specialists and we have been in the field for more than 25 years. We will be happy to forward a clients list to you. Please call my office to arrange for a meeting date.

Insert A

Insert A

For your reference,

ACTIVITY E WRITING PRACTICUM SENTENCE REWRITE FOR CLARITY

James Wolf is an independent distributor for the Indigo Vitamin Corporation. The following are the poorly written sentences from a letter he sent to prospective customers. Follow the instructions for correcting each sentence. Rewrite these sentences and put them into paragraphs. (Refer to Chapters 4 and 5 for review of active voice.)

Instructions

1. Use active voice; rearrange sentence
2. Concise writing; eliminate wordiness
3. Reverse order of sentence for a clear explanation of Indigo Vitamin System
4. Use active voice; rearrange sentence
5. Use introductory phrase in sentence for clarity
6. Rearrange sentence beginning with introductory clause
7. Rearrange order; start with "you" for emphasis
8. Change position of nonrestrictive clause in sentence
9. Emphasize "you"
10. Begin with introductory clause
11. Delete unnecessary words; change word order
12. Change to present tense; make stronger positive closing statement; add words as needed

1. Personal health and appearance are strongly emphasized by most people in their daily living.
2. However, schedules which are very busy don't always enable individuals to get the nutrients they need.
3. The combination of vitamins and minerals essential for a healthy body is the nutritionally balanced Indigo Vitamin System.
4. Furthermore, a weight-management program that includes daily exercise, proper diet, and stress control is encouraged by the Indigo Vitamin System.
5. I am so strongly committed to the Indigo Program after following it for one year that I have become an independent distributor of its nutritional line.
6. Please contact me if you wish to order any of the exceptional products with which you are impressed by the enclosed brochure.
7. An independent distributor could be you after using the products for a trial period.
8. As an independent distributor, which would give you a good income, you would represent an outstanding product line.
9. Items are discounted on all purchases you make personally.
10. You will receive the training support needed to provide customers with superior service whether you are a beginner or a veteran salesperson.
11. Health that is good is the key to a successful and productive life.
12. If you accepted our invitation, you can try the Indigo Vitamin Program that may change your lifestyle.

Bad News Letters

Two formats can be used to convey bad news: the direct or indirect approach.

In the **direct** approach, the reader is informed of the bad news immediately. The correspondence is courteous and direct. This approach is often used when the receiver needs to be informed in a direct, firm manner; for example, announcing a rate increase or rejecting a contract. It is not designed to promote goodwill.

The **indirect** approach delays the bad news or rejection by buffering it with a more positive opening and offering reasons for the news. The reader is then better prepared to receive the bad news. Use words that do not disturb the reader, but still deliver the message. You may also offer a suggestion or alternate action to soften the situation. This approach attempts to preserve customer goodwill; therefore, it is a more acceptable strategy.

Writing Pointers

Announcement of Bad News (Indirect Strategy)

- Open with positive statement about product, service, or customer.
- Provide objective information.
- State reason for change in policy or delay in delivery.
- De-emphasize the bad news.
- Offer alternative solution.
- Ask for cooperation or understanding.
- End with a positive statement; look forward to continued business.

Promotional Letter

- Capture reader's attention.
- Explain the sales campaign.
- State the duration of the offer.
- Offer some incentive for quick customer response.
- Indicate contact person, if possible.

Letter Format

- For letters addressed to companies, the salutation is either "Gentlemen" or "Ladies and Gentlemen."

Model Memorandum for Announcement of Bad News (Direct Approach)

TO: Sales Staff

FROM: Peter Lewis
Sales Director

DATE: (current date)

SUBJECT: Delay in the Delivery of Apex Notebook Computer

Indicates news directly and clearly to staff

Explains reasons for delay

The Apex Notebook Computer, a new addition to our line of computers that was targeted to be shipped on November 1, has been delayed. A slight modification in the design is causing this problem, and the new shipping date will be January 15. Unfortunately, we have lost the important holiday market.

Suggests alternatives

You should contact your accounts and explain the delay in shipment. Remind them that our policy is to offer the highest quality product; therefore, we had no alternative in the matter. Since this delay undoubtedly will have an adverse effect on your customers' sales, our firm is willing to offer a compromise. We shall give a 15 percent discount off the original contract price to compensate for any losses, and you may need to offer some incentives to keep the order.

Positive, upbeat closing

Demonstrates support and confidence in staff

I am confident that you will be able to rectify this matter with your clients and continue to maintain a good business relationship with them. A formal letter will be sent to our customers from this office. Feel free to call me for any further information.

ri

Model Letter of an Announcement of Bad News (Indirect Approach)

Date

MR ROY BENTLEY
CENTRAL COMPUTER STORE
100 CENTRAL AVENUE
EL PASO TX 79900-6375

Dear Mr. Bentley:

Opening positive statement about product — *Buffers bad news* — *Gives reason for delay*

Our firm prides itself on giving superior quality products to its customers. Unfortunately, a slight modification is required in the design of the Apex Notebook Computer, the latest addition to our line of quality computers. Therefore, our new shipping date is January 15. To maintain our firm's commitment to excellence in our products, we have no other alternative for this change in delivery date.

Demonstrates empathy — *Offers solution by compensating for loss*

We realize the delay comes before the holiday season, which is a crucial marketing period. To compensate for any losses, we are giving you a 15 percent discount on your order. Our sales representative will contact you personally to offer any assistance that you may need.

Positive closing

We apologize for this inconvenience; however, we are confident that as a loyal customer you understand the situation.

Sincerely yours,

Peter Lewis
Sales Director

ri

ACTIVITY F WRITING POWER: COMPOSING CORRESPONDENCE

CASE STUDY 1: BAD NEWS LETTER

Case: Prepare a letter to your customers informing them that there will be a 10 percent price increase on all printers as of the coming year. You need to explain that we are facing higher manufacturing costs as well as shipping fees. Remind them about our quality product. Use the indirect approach.

The form letter will be signed by Ms. Emily Prince, who is the Vice President of Sales for Arrow Electronic Corporation, 115 Midland Drive, Toledo, OH 43600–2432.

Organize Your Ideas

1. What positive statement would you make in your opening sentence?

   ____

2. What type of statement would you make preceding the announcement of a price increase?

   ____

3. How would you convince your customers that this price increase was the lowest amount the company could assess and still maintain a quality product?

   ____

4. What would you say in the closing statement to keep the customer's loyalty?

   ____

CASE STUDY 2: BAD NEWS INFORMATIONAL LETTER

After reading the following case, respond to the questions, which will help you organize your ideas before beginning to compose the bad news letter. Then draft the letter. Use the indirect approach.

Case: This form letter will be sent to Computer-Op customers signed by Richard Thompson, President. Computer-Op is announcing to customers that it plans to close all the branch stores on December 31 (current year). Explain that the reason for the closings is the result of higher costs in employee benefits, rent, and supplies. Management attempted to cut costs by reducing inventory and operating expenses. They did not want to decrease services to their customers. Computer-Op has always offered quality products at moderate prices. Therefore, the firm made the decision to close some of its stores. The main store at One Park Terrace will remain open and will carry a complete line of computers and office equipment.

Organize Your Ideas

1. What is the most important point you want to make?

 ____

 ____

 ____

2. What would you say in the opening sentence?

 ____

 ____

 ____

3. Where in the letter would you mention the bad news of store closings?

 ____

 ____

4. What is your alternative solution?

 ____

 ____

 ____

5. What would you say in the closing statement to end on a positive tone?

 ____

 ____

 ____

CASE STUDY 3: MEMORANDUM TO SALES STAFF

After reading the case below, draft the memorandum.

Case: Your general manager, Walter Perks, asked you to prepare a memorandum to all sales representatives. It should inform them about the annual three-day sales meeting. This is to be held at the Midway Hotel in Los Angeles, California, from April 1 to 3. The 15 percent decrease in sales this past year is a major concern of management, and it will be the major topic discussed. What are the reasons for the decline and what is the impact on the firm? The sales representatives will be expected to talk about sales figures for the year, projected income for the next year, and plans to attract new customers. They will be asked to give input on ways to improve company service and product line.

CASE STUDY 4: PROMOTIONAL LETTER

Read Case Study 4, develop an outline to guide you in writing, then compose the letter.

Case: You are working for a new car dealership in your local community, and your employer has given you the task of composing a promotional letter to be sent to prospective buyers. Discuss the line of cars you are promoting, highlight special features, and offer customers an incentive. Appeal to customers' emotions and interests. You may want to refer to automobile ads in your local newspaper for ideas.

CASE STUDY 5: PROMOTIONAL LETTER

Read the case study; then compose the letter.

Case: Mr. Larry Torres, President of Peabody Communication, Inc., is writing a promotional letter to Mr. Richard Deveroux, Vice President of Public Relations, ABC Products, One River Road, Boston, MA 02241. The letter describes the services of Peabody Communications, Inc., a marketing consultant firm specializing in media communications. This company custom-designs marketing plans for both small businesses and multinational corporations. Staff members are experienced in marketing research. They study the target market that is suited for each product line. The analysts then propose a comprehensive package of online marketing and print media to promote the brand of products. They handle the entire promotional campaign. Mr. Torres invites Richard Deveroux to discuss a successful communication program for ABC Products.

CASE STUDY 6: E-MAIL

Draft an e-mail to Jose Rodriguez (jose@electronicgraphics.com) from Mark Price, Vice President of Operations, informing him that the project to install a web page for Loral Products has been postponed. Communicate to Jose that his proposed web site was impressive. The graphics were creative. Mark Price liked the idea of including a column on health information that would be a drawing card for customers. However, explain to Jose that there has been a slowdown in sales. Even with a projected net sales increase of 20 percent a year, the budget director at this time has placed a freeze on all new projects. Therefore, the plan to implement a web site will be delayed. As soon as there is an improvement in the economy, the firm intends to initiate a web page. Tip: Use a closing sentence that offers a positive solution.

Include your own subject line.

ACTIVITY G REPORT WRITING POWER: EXTERNAL BUSINESS PROPOSAL

An **external business proposal** is prepared by a firm or an individual outside a company. It is more formal than a memo proposal, which is prepared internally. The external business proposal contains many of the same parts of a formal report. Projected costs are generally included. It does not include an index or bibliography that is usually found in a formal report. The executive summary is optional.

A **transmittal letter** frequently accompanies the business proposal. This letter introduces the reader to the text of the proposal. Include a brief statement that describes the objective of the proposal and persuades the reader to take action. End the letter with a request for approval and offer assistance.

One of the main purposes of the proposal is to sell an idea; therefore, the text should reflect this goal.

Writing Pointers for External Business Proposals

Components of an external business proposal are presented in a formal structure.

- Letter of transmittal
- Cover page
- Table of contents
- Body

  Summary—overview of plan

  Purpose—goal of proposal

  Need or problem—discuss situation

  Benefits of proposal

  Description or design of plan

  Projected costs—be as specific as possible
- Conclusions or recommendations

Model Transmittal Letter to Accompany an External Business Proposal

Current Date

MS DEBRA MIRANDA
PRESIDENT AND CEO
RENAISSANCE DEPARTMENT STORE
109 BLOSSOM DRIVE
SAVANNAH GA 31400-3522

Dear Ms. Miranda:

Highlights important aspects of proposal

As you requested, enclosed is the "Proposal for the Expansion of the Renaissance Department Store's Web site." The online retail business has become increasingly popular among shoppers who have limited free time. Studies show that online shopping will increase revenues for retailers.

Persuades reader to take action

Our proposal presents a positive plan that will result in added sales for Renaissance. The present web page is inadequate to meet the needs of the consumer. Redesigning the web page, which will include a personal shopper service, will enable Renaissance to remain competitive in retailing.

Requests approval

With your approval, we will be able to implement the plan immediately. If you have any questions, please call me.

Sincerely yours,

Joel VanderMeer
Retail Marketing Consultant

ri

Enclosure

Model Business Proposal

TITLE PAGE

PROPOSAL FOR THE EXPANSION
OF THE
RENAISSANCE DEPARTMENT STORE'S WEB SITE

Submitted to
Debra Miranda, President and CEO
Renaissance Department Store
Savannah, GA 31400-3522

Submitted by
Joel VanderMeer
Retail Marketing Consultant

Current Date

CONTENTS

Model External Business Proposal

PROPOSAL FOR THE EXPANSION
of the
RENAISSANCE DEPARTMENT STORE'S WEB SITE

SUMMARY

Gives overview of plan

Online retail selling is a major source of revenue for department stores. The purpose of this proposal is to redesign Renaissance's web site so that it offers expanded services. The prime focus of the web site will be the personal shopper service.

PURPOSE

Presents goal and importance of proposal

The purpose of this proposal is to expand the Renaissance Department Store's web site so that it will enhance customer service and increase the market share of online buying. Since the Internet plays a major role in retail selling, the firm's web page must not only attract shoppers but also appeal to repeat customers. To achieve this objective, the web page will include a personal shopper feature.

PROBLEM

Discusses current situation

The present web site is basically an online catalog that features the store brand clothing line, cosmetics, and accessories. Online purchasing is not a dominant part of the firm's present marketing policy. It accounts for only 10 percent of total sales.

States problem or need clearly

In comparison with other retailers' web sites, it is apparent that competitors have the advantage. They feature free gifts, periodic sales, free delivery, and customer discounts. Information on the latest fashions are frequently given on the web sites as well as tips on budgeting. The web page is used by competitors as an integral part of their marketing program.

BENEFITS OF THE PROPOSAL

The proposed addition of a personal shopper service to Renaissance's online shopping offers hassle-free shopping

Stresses benefits that influence decisions

for busy people. The human contact of a personal shopper makes the shopping experience more individual. Experienced personnel who are familiar with the latest clothing line and accessories will assist the customer in making a purchase without wasting time in the store. Customers will be more satisfied with their purchases, thus lowering the percentage of returned merchandise.

Sales would increase because shopping is easier and less stressful. A more positive image of Renaissance's Department Store would develop as more unique services are provided.

DESCRIPTION OF THE PROPOSAL

Describes design of proposal and task

When customers make online purchases, they will be referred to a personal shopper. This employee, who is a knowledgeable salesperson, will assist the shopper in making wise decisions. Each customer who uses Renaissance's web site will have a personal profile outlining the shopper's clothing size, style, color preferences, and previous purchases.

The web page might also include other information about merchandise; for example, suggestions on clothing accessories and wardrobe coordination, which would be of interest to shoppers, and beauty questions and answers related to cosmetics, which would appeal to women clients. This added service expands the web site from a catalog approach to an informative fashion magazine.

Initially, the present staff of personal shoppers would handle this new service. They are the best qualified individuals to introduce the program because of their experience. Additional staff will be hired as the need arises.

PROJECTED COSTS

Indicates costs

The proposed changes would not be costly because the present staff, with assistance from the consultant, could redesign the home page. Software development to update the web site, however, would cost approximately $25,000.

As an incentive to increase sales, a bonus program for the personal shoppers could be introduced, such as awarding a sum of $10 for every customer who purchases more than $500 annually.

A 20 percent increase in sales is anticipated during the first year of the proposed plan. This increase may result in the hiring of new personnel at the salary of $30,000 that is presently paid to a personal shopper.

The consultant's fee to oversee the implementation of the plan will be $10,000.

CONCLUSION

Brief statement summarizing proposal

Redesigning the web site will undoubtedly increase the number of customers who use online shopping, thus creating more business for Renaissance. The web is a key element in enabling Renaissance to remain competitive in retail marketing.

CASE STUDY: EXTERNAL BUSINESS PROPOSAL

In this chapter, you are required to prepare an external business proposal. Follow the Writing Pointers that appear at the beginning of this chapter. You will find a model of an external business proposal in this chapter to use as a guide. The following are suggested topics you may use in developing the proposal. However, feel free to write a proposal on a topic of interest to you or with which you are familiar.

You will need to research the project to support your statements and plan of action. Be specific and provide details of the design and implementation of the plan.

Suggestions

- Imagine that you are a consultant who has been hired by your college to submit a proposal on one of the following topics: improvement of procedures, development of curriculum, expansion of college offerings, updating equipment, or establishing programs that will reach out to the community.
- Think of a procedure that you would like to introduce or change at your place of work. Assume that you are an outside consultant when writing the proposal.

ASSESSMENT: CHAPTER 9

English Power

Directions: *Insert an apostrophe or 's where necessary and key each sentence.*

1. The firm charges its customers a 5 percent shipper fee.
2. The new organization, Womens Business Alliance, will meet to discuss ways to start a small business.
3. Taylor and Smith sporting goods shop has introduced a mail-order department.
4. The company sales campaign is directed to a multicultural market.
5. Our in-law anniversary gift was a certificate to Appleby Boutique.
6. There are many types of salespeople; for example, manufacturers representatives who deal only with retail stores.
7. To raise capital, management decision is to issue stock.
8. The competitor strategy for marketing toys during this holiday season is similar to ours.
9. I'm not sure what bonus will be given for Travis outstanding sales performance.
10. The advertising agency contract included the creation of a slogan for the sponsor product.

Editing Power

Directions: *Edit, format, and proofread the following unarranged memorandum. Supply subject line, other missing parts, and paragraphs; key it correctly. Send this memorandum to sales representatives from Donna King, Vice President.*

```
I am pleased to announce that Paula Brown is joining our
companys staff as regional sales manager. This new
position as you know was created to handle increased
sales responsibilities and Paula will be responsible for
the Northeast region. Paula comes to us with an
extensive background in general sales telemarketing and
computer programming. Her latest accomplishment was to
implement a high-tech communication system for the sales
force at Apex Corporation a billion-dollar business.
Paulas professionalism and expertise will contribute
greatly to our organization. Paula and her husband
Robert will be relocating to Stamford CT where she will
be joining our office on July 1.
```

Writing Power

Directions: *After reading the case below, draft a memorandum.*

As office manager, you are to notify your sales staff that the fourth-quarter earnings were down by 15 percent. This downward trend has continued for the past year; therefore, certain budgetary changes are necessary. A freeze on salary raises will go into effect beginning next month, and sales commissions will be lowered to 2 percent of total sales. All travel expenses exceeding $500 must be approved by you. Inform the sales staff that this is only a temporary arrangement and that the company will reinstate previous policies as soon as sales begin to show an upward trend. Encourage the staff to cooperate during these difficult times.

> "No problem is insurmountable. With a little courage, teamwork, and determination, a person can overcome anything."
>
> —*B. Dodge*

10

Traveling with Colons and Pronouns

WHAT YOU WILL LEARN IN THIS CHAPTER

- To apply correct pronoun usage
- To apply correct colon usage
- To develop a business vocabulary in **TRAVEL AND LEISURE**
- To proofread and edit correspondence and reports
- To compose a memorandum
- To compose a response to a request
- To compose a letter of notification
- To compose a claims letter
- To compose a response to a claim
- To write an e-mail
- To prepare a formal report

Corporate Snapshot

SOUTHWEST AIRLINES. In the 1970s, a dream was born. A little three-jet airline started business in the Southwest. Its mission was to offer the highest quality consumer service. Southwest Airlines pledged to provide safe, affordable, reliable, timely, and courteous air travel. The skeptics doubted that a small regional airline that offered low fares could compete against the industry giants.

Thirty years later, Southwest has proved to the industry that its way of doing business is successful. Today it is one of the largest domestic commercial airlines and has remained faithful to its pledge of commitment to the customer. The airline pioneered senior discounts and ticketless travel. The fleet of planes is new and roomy with an uplifting interior decor giving a message that flying is fun.

Southwest has become a model for many of the new discount airlines.

Source: *http://www.southwest.com/about_swa*

ACTIVITY A LANGUAGE ARTS RECALL

Colons

Colons appear at the end of an independent clause when followed by a list or an explanation.

Rule 10.1: Place a *colon* after the phrase *as follows*, the *following*, or *as detailed below* after an independent clause that is followed by a list or an explanation.

Example: Travelers are interested in touring ***the following countries:*** England, Spain, and Australia.

Rule 10.2: Place a colon between the hours and minutes when the time is expressed in figures.

Example: The plane leaves at ***1:30*** p.m.

Rule 10.3: Use a colon after an independent clause that introduces a quotation of three or more lines.

Example: Here is the advice given to travelers in the new edition of *The Travel Guide:*

> Charter flights are often the least expensive way to travel on transcontinental routes. A traveler must be willing to adjust to the limited flight schedule of the chartered airline.

Rule 10.4: Place a *colon* after a salutation in business correspondence.

Example: ***Dear Ms. Johnson:***

Pronouns

A **pronoun is** is a word that is used to replace a noun. "Who" and "whom" are pronouns that refer to people. They may be singular or plural in number and may be interrogative or relative.

An **interrogative pronoun** is used in a sentence that asks a question. A **relative pronoun** introduces a clause within a sentence that modifies a noun.

Note: To help you determine whether to use *who* or *whom*, rearrange the sentence and substitute a personal pronoun for *who* or *whom*.

- To determine if the pronoun *who* should be used, substitute *I*, *we*, *she*, *he*, or *they*.
- To determine if the pronoun *whom* should be used, substitute *me*, *us*, *her*, *him*, or *them*.

The different forms of personal pronouns are listed in the chart below:

Subject Pronouns	Object Pronouns	Possessive Pronouns
I, we	me, us	my, mine, our, ours
you	you	your, yours
he, she, it,	him, her, it	his, her, hers, its
they	them	their, theirs
who, whoever	whom, whomever	whose

Rule 10.5: The pronoun *who* is used in the subjective case.

Example: ***Who*** signed up for the trip to the Canadian Rockies?

Check: She signed up for the trip to the Canadian Rockies. (She is the substitute for the pronoun who.)

Rule 10.6: The pronoun *whom* is used in the objective case.

Example: ***Whom*** did you sign up for the trip to the Canadian Rockies?
Check: You signed ***him*** up for the trip to the Canadian Rockies. (*Him* is the substitute for the pronoun *whom*.)

A relative pronoun introduces a clause that modifies a noun. The way the pronoun is used in its own clause determines if it is *who* or *whom*. For example, a pronoun that functions as a subject is in the subjective, or nominative, case (who); a pronoun that functions as an object is in the objective case (whom).

Rule 10.7: The pronoun *who* is used as the *subject of a verb in a clause*.

modifies

Example: Blanca is the client ***who*** is planning a trip to South America. (*Who* the subject of the verb *is planning*. *Who* relates to the word *client*.)

Check: ***The client*** is planning a trip . . . .

She is planning a trip to . . . . (*She* substitutes for the pronoun *who*.)

Rule 10.8: The pronoun *whom* is used as the object of a verb and a preposition.

Example: ***Whom*** did you travel with to the Far East? (Rearrange order of sentence.)

Check: You traveled with ***him*** to the Far East. (object of preposition *with*.)

Example: The applicant ***whom*** I interviewed for the position was willing to work overtime.

Check: I interviewed ***her*** for the position . . . . (*Her* substitutes for the pronoun *whom*.) (*Whom* is the object of the verb *interviewed*.)

Rule 10.9: *That* and *which* are relative pronouns. *Which* refers to things; *that* refers to people or things. Always use *which* when introducing a nonrestrictive non-essential clause; *that* is generally used to introduce a restrictive essential *clause*.

Examples: The European tour, ***which*** was scheduled for next month, was canceled. (nonrestrictive)

The island ***that*** he visited was beautiful. (restrictive)

Check for Understanding

Directions: For each of the following sentences, indicate the reason for using *who* or *whom.*

1. *Who* will plan the itinerary for the club members?

   Reason: ______________________________

2. *Whom* did you write to for a travel guide?

   Reason: ______________________________

3. The frequent flyer mileage credits will be given to those travelers *who* accumulate 30,000 miles.

   Reason: ______________________________

4. The tour director *whom* we have used in the past is an expert in archaeology.

   Reason: ______________________________

5. The screening committee was to recommend three candidates from *whomever* they interviewed.

   Reason: ______________________________

(Answers are below.)

Answers to "Check for Understanding"

1. subject of verb
2. object of preposition
3. subject of verb
4. object of verb
5. object of verb *interviewed*

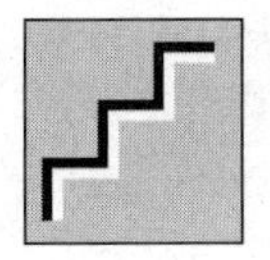

Building English Power: Colons and Pronouns

Exercise 1. Who and Whom, That and Which For each of the sentences that follow, select the correct form of the relative pronoun in parentheses and write it in the space provided.

______________ 1. We are offering a discount to members (who, whom) renew their subscriptions before April 1. (10.5, 10.7)

_______________ 2. To (who, whom) do we mail the check for the tickets? (10.6, 10.8)

_______________ 3. It is not everyone (who, whom) can qualify for membership to the travel club. (10.5, 10.7)

_______________ 4. The young lady (who, whom) I met at the country inn was the editor of the travel magazine. (10.8)

_______________ 5. The couple (who, whom) I met at the hotel is from your home town. (10.6, 10.8)

_______________ 6. (Whoever, Whomever) they appoint director of the travel agency will have five assistants (who, whom) will report to him or her. (10.5, 10.6, 10.7, 10.8)

_______________ 7. (Who, Whom) shall we consider as our representative to the trade show? (10.6, 10.8)

_______________ 8. The travel guide, (which, that) I purchased yesterday, has detailed city maps. (10.9)

_______________ 9. We are interviewing ten applicants, many of (who, whom) have more than five years' experience. (10.6, 10.8)

_______________ 10. The suggested itinerary for the tour will be planned by (whoever, whomever) you select. (10.6, 10.8)

_______________ 11. The new travel agent (who, whom) I met today came from Florida. (10.6, 10.8)

_______________ 12. I do not remember to (who, whom) I sent my trip map. (10.6, 10.8)

_______________ 13. The travel club members (who, whom) have paid their dues are entitled to a free vacation kit. (10.5, 10.7)

_______________ 14. He is one of the tourists (who, whom), I believe, lost his luggage at the airport. (10.5, 10.7)

_______________ 15. Does anyone know (who, whom) is planning to attend the travel agent's seminar? (10.5, 10.7)

Exercise 2. Who and Whom, Colon, and Other Punctuation Correct and key the following sentences. Be sure to express time in figures, using a.m. or p.m.

1. We plan to leave for the airport at ten past three in the afternoon with (whoever, whomever) is available at that time. (10.2, 10.5, 10.7)
2. (Who, Whom) is interested in touring the following cities London Milan and Athens? (10.5, 10.7, 10.1, 7.1)
3. Tourists (who, whom) are planning to travel by car should review the following checklist get car tuneup obtain roadmaps get sightseeing information and carry emergency supplies. (10.5, 10.7, 10.1, 7.1)
4. The manufacturing executives will visit their plants as follows Bangkok Thailand Jakarta Indonesia and Kobe Japan. (10.1, 8.3)
5. (Who, Whom) has scheduled the afternoon tour for one fifteen? (10.5, 10.7, 10.2)

ACTIVITY B BUILDING WORD POWER

The words below are commonly used in the field of travel and leisure. From the following list, select the word or phrase that correctly completes each statement. Refer to the Personal Dictionary, if necessary.

accrues
budget
cost conscious
diversified
emerging
extended-stay hotels
globalization
instantaneously
itinerary
liaison
no-frills
nominal
preliminary
priority
segments
strategies

______________ 1. As lifestyles change, different types of vacation packages are ________ to meet consumer demands.

______________ 2. When planning a tour, be sure to have a(n) ________ that outlines in detail your accommodations, special events, and daily activities.

______________ 3. A(n) ________ that projects spending and income is an important aspect of managing money.

_______________ 4. Many people are sensitive to the price of items. These individuals are described as being _________.

_______________ 5. For a(n) _________ fee, which is relatively low, you can purchase liability insurance when renting a car.

_______________ 6. There are travel agencies that specialize in targeting certain _________ of the market.

_______________ 7. A successful vacation requires _________ planning before taking the trip.

_______________ 8. The safety of an area is a major _________ in choosing a vacation site.

_______________ 9. With today's fax machines, information is transmitted _________; and the message is received within minutes.

_______________ 10. There are _________ ways to travel so that people have many choices when planning a trip.

_______________ 11. _________ is a term often used to describe businesses that interact and work jointly with foreign firms.

_______________ 12. A(n) _________ refers to an individual who brings people and organizations together for a special purpose.

_______________ 13. Under the frequent flyer plan, a traveler who _________ the prescribed minimum number of miles can receive a free ticket.

_______________ 14. To improve sales, businesses develop different _________ or plans to reach their goals.

_______________ 15. The new low-fare, _________ airlines do not offer passengers free meals.

_______________ 16. If families prefer more homelike accommodations for a few weeks, they should consider _________.

On the Alert

- an (indefinite article)—used before words beginning with vowels and before silent "h."

  Examples: **An** ideal city; *an* honorable man

- chose (verb, past tense)—selected

  Example: They ***chose*** a travel package to Hawaii.

- choose (verb, present tense)—to select

  Example: Many Americans ***choose*** the national parks for their vacation.

- Hyphenate compound adjectives that precede a noun.
  Example: ***toll-free*** number
- Do not add a decimal point or zeros to whole dollar amounts.
  Example: **$40**
- Use the dollar symbol ($) when expressing exact amounts of money; do not write out the word "dollars" when the dollar symbol is used.
  Example: **$50,000**

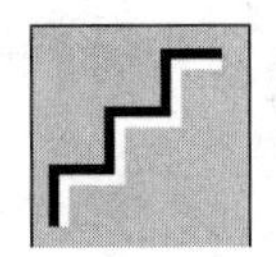

ACTIVITY C INDUSTRY PERSPECTIVES: TRAVEL INDUSTRY

Directions: Using appropriate revision marks, proofread and key the following report. The title is EMERGING TRENDS IN THE TRAVEL INDUSTRY

Rule Reference

6.3

7.1

3.4

The travel industry is greatly affected by the economy and the world situation. In recent times the uncertainty of the marketplace his increased competition within the industry. Companies compete for the consumer dollar by offering discount prices, travel perks and attractive vacation packages. Everyone recognize that the Internet has become a very important means of doing business.

CHANGES IN MARKETING STRATEGIES

7.1

Marketing strategies have shifted to accommodate the globalization of our economy, the diversified population from countries worldwide and the changing demands of consumers. Airlines and hotels attempt to accomodate the client who needs to conduct business and keep in touch with the home office. Excess to high speed computers and fax machines are readily available.

7.1 Hotels, airlines, car rental agencies and tour companies use the Internet to offer attractive programs

10.7 for the leisure traveler whom may be interested in low-cost package deals.

7.7 As competition remains high, low-fare no-frills airlines have become popular among vacation travelers. Some hotel chains offer free breakfasts and snacks to

10.7 attract patrons whom prefer this type of convenience.

10.9 Extended-stay hotels now offer suites which accommodate

9.1, 10.5 families. In todays market, whomever offers the best deal will get the business.

CHANGES AFFECTING THE TRAVEL BUSINESS

In the travel industry, the following are some of the emerging trends that are affecting the way business is conducted.

10.3

The Internet has become a favorable way of booking air fights hotel accommodations and car rentals.

7.1

Many airlines and hotels use the Internet to offer discount rates.

- Increased choices of discount arilines offer no-frill packages to travelers.

10.7 ■ Cruises have become popular for individuals whom enjoy a sea vacation with entertainment.

3.6 ■ A increased number of travelers enjoys touring historic places or specific geographic areas of interest.

■ The changes that are occuring in the travel industry will no doubt benefit the traveling consumer who wants quality service at the best price.

Did you find these errors?

8 commas	2 subject/verb agreement
1 colon	1 word usage
4 pronouns	1 article
1 possessive	2 typographical

ACTIVITY D EDITING POWER

Task A: Edit, format, and proofread the following unarranged form letter. Use proofreaders' symbols to mark the errors, then key it correctly.

Use "Dear Customer:" as the salutation. The letter concerns additional travel privileges, and it is written by Edward Lawson, General Manager.

Errors
- pronouns
- commas
- spelling
- subject-verb agreement
- capitalization
- hyphen
- typographical

In an effort to offer our customers additional travel priviledges we are pleased to announce our new frequent flyer plan with Executive Airlines. Whenever you arrange a tour or make plane reservations through us you will recieve bonus air miles based on the number of miles flown and type of land arrangements. Anyone whom accrues 25,000 miles can apply for a free round trip ticket in business class to any U.S. city served by executive

```
Airlines. If you have a minimum of 10,000 miles you will
be able to get a cash discjount on one of our regularly
priced tickets. This new policy offers our customers a
greater insentive to take more vacations. An added
amenity for our customers is our new priority
reservations system. This computer4ized service allows
you to call our toll free number for reservations. Our
representatives are on call seven days a week for
vacation information and suggested travel packages.
Whomever wishes to take advantage of this frequent flyer
plan need only to fill out the enclosed application
form. There is no charge to enroll in the frequent flyer
plan. Why not join the thousands whom have already taken
advantage of this offer.
```

Task B: Edit, format, and proofread the following rough-draft letter for all errors; key it correctly. Address this letter to Ms. Colleen Burke, National Retailing Association, 100 Pine Street, Pittsburgh, PA 15218-7622. This letter is from Mr. Arnold Black, Director.

```
In response to you letter of June 15 we are pleased that
you are intrested in holding the National retailing
                          (South Carolina)
Convention in Charleston. Our staff is ready to assisat
you wil arrangement from hotel availability to
conference facilitites. Charleston is a ideal city to hold a
                 It
convention. Charleston is more than a historical sight.
```

Charleston
~~It~~ offers a variety of attraction, such as cultural events sports and fine beaches. Our staff is knowledgable in arranging all ~~any~~ types of meetings. We will ~~also~~ coordinate transportation to and from the airport and shuttle services with in the city. Mr. Gary Scott whom is our liaison coordinator will work with your group in planning the convention. ~~all arrangements~~ He will ~~can also~~ make suggestions for group tours entertainment and local events. ~~the most efficient ways to facilitate your meetings.~~ Mr. Scott will contact you next week for preliminary discussion. Thank you for chosing charleston as your convention city; we are delighted to have you with us and hope to work with you in the near future.

ACTIVITY E WRITING PRACTICUM

EXERCISE 1 POINTING AND LIMITING ADJECTIVES

The adjectives *this, that, these,* and *those* point out a noun and indicate whether the noun is singular or plural. The adjective must agree with the noun. To be an adjective, it must be followed by a noun; otherwise, it is a pronoun.

Examples: ***This*** travel agency is rated high. (adjective, singular)
Those maps are too old to use. (adjective, plural)
These are the tools I need. (pronoun, plural)
That is the proper route. (pronoun, singular)

Directions: Rewrite each of the following sentences so that the correct pointing adjective is used. You might have to change the pronoun or adjective, add an "s" to the noun, or eliminate a word.

Example:

Incorrect: Them there cruise ships leave for foreign ports daily.
Change: Change "them there" to "those."
Correct: Those cruise ships leave for foreign ports daily.

1. Those kind of package tours are inexpensive.

   ____

2. That there tour guide was well informed.

   ____

3. That cruises cater to senior citizens.

   ____

4. Please ask them there travel agents to come to the tourism convention.

   ____

   ____

5. Are these type of flights eligible for frequent flier miles?

   ____

6. This here brochure describes a very exciting trip to the Galapagos Islands.

   ____

   ____

7. I can't decide between this accommodations and that there luxury rooms.

   ____

   ____

8. That countries which are in the news because of terrorism show a sudden drop in tourism.

   ____

   ____

9. This bellhop over there took such good care of my luggage that I gave him a big tip.

   ____

   ____

10. These kind of side trips are available for nominal fees after you pay for the total package.

____

____

EXERCISE 2 WRITING SENTENCES FOR CONCISENESS AND CLARITY

Concise sentences use only words that are essential to the meaning. Avoid unnecessary words and those that are repetitious (redundant). Generally, wordiness hides ideas; conciseness promotes clarity.

Directions: Revise the following sentences to eliminate wordiness.

1. At this point in time, people prefer shorter vacations.
2. Because of the fact that there was an increase in the airfare, we decided to take an auto trip.
3. In the event of a strike, the airline will offer passengers a refund.
4. For the most part, individuals still prefer to discuss vacation plans with a travel agent.
5. As per your request, the travel brochure for Alaska has been mailed to you under separate cover.
6. We acknowledge receipt of your payment for your European vacation tour.
7. I am writing this letter to tell you that I am unable to attend the seminar next week.
8. I wish to thank you for your support and help.
9. In spite of the fact that we have sent you numerous reminders, we have not received your check for the overdue balance in the amount of $500.
10. We are sending you our check in the amount of $100 in order to reserve the meeting room.

EXERCISE 3 WRITING PARAGRAPHS FOR CONCISENESS AND CLARITY

Read the following two paragraphs to assess their conciseness and clarity. Using proofreaders' symbols, delete unnecessary words. Key each paragraph.

Paragraph 1

Are you planning to take a trip and vacation this summer? Now we are happy to tell you that there is a fast, efficient way of doing this. Trip Saver is a new software package that will without a doubt help you plan your trip and make all arrangements. This package, believe it or not, will map out and indicate a road and route, give you suggested side trips, and give you an approximate estimated driving time. This software program will also offer overnight accommodations and lodgings for you to stay at. If you want a quick, fast, short, and scenic way to travel, use Trip Saver.

Paragraph 2

We are very pleased to let you know that if you join our club, Five Star Travel Club, you will receive an opportunity to take a vacation at the low price of $199 for each member of your family.

Our club, Five Star Travel Club, offers its members low and greatly reduced rates in airfare, hotel accommodations and good sleeping quarters, and car rentals. Our destinations include exotic Hawaiian resorts to cosmopolitan U.S. cities.

To be eligible for these inexpensive, cheap vacations, all you have to do is to join our club today. Fill out the enclosed card, and we will bill you for the $25 membership fee.

Writing Pointers

Response to Special Request Letters

- Convey to readers your thorough knowledge of the topic.
- Express enthusiasm about the topic.

- Emphasize important features.
- Enclose literature, if available.

Letter of Notification to Customer

- Notify customer of situation (changes in price, service, or policies).
- Open with positive statement if news is bad.
- Include benefits of product or service.
- Indicate reasons for changes.
- Express appreciation for past business.
- Suggest alternatives.
- Maintain goodwill.

Claims Letter

- Be specific in your claim.
- Give a clear explanation of the problem.
- Provide specifics: dates, costs, locations.
- Appeal to the firm's reputation of fairness.
- Suggest a solution or adjustment.
- Use a courteous closing.

Letter of Response to Claims

- In the opening sentence, deliver the good news or positive response to claim.
- In the body of the letter, explain the problem and/or give a solution.
- Use a pleasant closing; express confidence in future business relations.

Model Letter of Notification

Date

MR AND MRS LAURENCE CHARLES
3641 MAGNOLIA BOULEVARD
SAVANNAH GA 31400-4732

Positive opening

Dear Mr. and Mrs. Charles:

B & E Tours has enjoyed a wonderful reputation in providing exciting tours to out-of-the-way places. Travelers who enroll in our tours have memorable experiences that remain with them forever.

Notify customer about change

We are disappointed to inform you that the High Road to the British Isles and the Treasures of the Amazon tours have been canceled this year. Interest in these two tours has been minimal, and they could not be offered at this time.

Reason for change

Suggest alternative

We know you have expressed an interest in these two tours; therefore, we have enclosed a brochure with other possible itineraries you may find interesting. We hope to hear from you soon so that we can help plan your next exciting trip.

Maintain goodwill

Sincerely yours,

Julia Fisher
Tour Coordinator

ri

Model Claims Letter

Date

MR. WAYNE REED
MANAGER
LEISURE TOURS, INC.
3000 NORTHERN PARKWAY
PHILADELPHIA PA 19100-3200

Dear Mr. Reed:

Positive opening — I have been looking forward to the spring tour to Ireland planned by your agency. It included all the places that I wished to visit.

Explanation of problem — Last week I was informed that I need to undergo major back surgery. The recovery period will be lengthy. Therefore, I need to cancel my booking with you.

Appeal to firm's reputation — I am aware that the deadline for a refund has passed. Your travel firm has a fine reputation in accommodating its clients. In view of this unforeseen serious situation, I

Solution — hope you will consider refunding to me the full amount of $2,000.

Courteous closing — Please let me know what action your firm will take.

Sincerely yours,

John McCulley

ri

ACTIVITY F WRITING POWER: COMPOSING CORRESPONDENCE

CASE STUDY 1 MEMORANDUM TO STAFF

After reading the following case, respond to the questions that follow, which will help you organize your ideas before beginning to compose the memorandum; then draft it.

Case: Linda Billings, sales manager, has been asked to notify her staff of the revised travel expense policy, which becomes effective immediately. All travel arrangements will be made through XYZ Travel Agency. Hotel accommodations must be booked at mid-range rates, airline reservations cannot go beyond business rate, and meal expenses cannot exceed $50 per day. Indicate that these restrictions are being made to reduce expenses.

Organize Your Ideas

1. What is the subject of the memorandum?

____

2. What would you include in the opening sentence?

____

____

3. What specific details would you include in the body of the memorandum?

____

____

____

4. What would you say in your closing statement?

____

____

CASE STUDY 2 LETTER OF RESPONSE

Read the case below, then answer the questions that follow.

Case: As an employee of the local Convention and Visitors Bureau, you receive a letter from Mr. and Mrs. Joseph Barnes at 14 Sterling Drive, Stamford, CT 06900–3412, requesting information about your town or city. Send a response highlighting the interesting points about your town or city. You may want to include popular sites, historical places, festivals, sports, and cultural events. You should also enclose a brochure that describes in detail the town's hotel accommodations, restaurants, tours, and transportation facilities. Warmly welcome the Barneses to your city.

Organize Your Ideas

1. What is the most important point in responding to the inquiry?

   ____

   ____

2. What would you include in the opening statement?

   ____

   ____

3. What specific details would you include in the body of the letter?

   ____

   ____

   ____

4. What would you say in the closing statement to encourage tourism to your city?

   ____

   ____

CASE STUDY 3 FORM LETTER OF NOTIFICATION

Read the case, and compose the letter.

Case: Compose a form letter to be sent to those individuals whose subscriptions to *Vacation Get-A-Way* magazine are ending. There is an increase in the subscription rate to $24 for 12 issues. The rate increase is due to higher costs of paper and production. This rate is still 50 percent lower than the newsstand price. Use the indirect approach to give the bad news of a rate increase by mentioning the positive features of the magazine. Subscribers are excited about the feature columns that are written by experts in the field. Informative articles discuss foreign vacation cities, cruise tours, and scenic worldwide places. In future issues, a food and shopping section will be included. To renew, use the postage-paid form. The salutation should read Dear ______________: (the subscriber's list will be used to address individual names). Lara Hung, Managing Editor, will sign the letter.

CASE STUDY 4 BAD NEWS CLAIMS LETTER

Read the case; then compose the letter

Case: Write a letter on behalf of Sara Hong to Customer Services, Express Airlines, LaGuardia Airport, New York, NY 11105. On May 1 (current year) Sara Hong flew from Chicago to New York City on Express Airlines flight 630. Her luggage was badly damaged. It was ripped on the side and could not be repaired. This was a new 23-inch piece of Sports brand luggage; the retail price was $125. Sara would like to be reimbursed the full amount for the damage so that she can replace this piece of luggage.

CASE STUDY 5 RESPONSE TO A CLAIMS LETTER

Read the case; then compose the letter

Case: Compose a letter in response to Ms. Sara Hong's claim for reimbursement of $125 to cover the cost of her damaged luggage. Express

Airlines will reimburse her for the damages. The company is implementing new baggage handling procedures so that such problems will not occur in the future. Mention to Ms. Hong that Express Airlines is committed to satisfy its customers. End on a positive note expressing confidence in serving her in the future. Address the letter from Ted Wells, customer advisor, to Ms. Sara Hong at 550 Thornhill Road, Great Neck, NY 11420-1834.

CASE STUDY 6 E-MAIL

Read this case; then complete the e-mail to Jim Lange.

Case: Send an e-mail to Jim Lange (jiml@yahoo.com) notifying him that the cruise tour on September 1 (current year) through the Panama Canal has been canceled because of a lack of bookings. Suggest an alternate tour of the Caribbean ports or a refund of his $500 deposit. Express the apologies of Leisure Tours, Inc.

ACTIVITY G REPORT WRITING POWER: FORMAL REPORTS

Formal reports usually deal with complex issues that may involve research and analysis of data. In business, it may be necessary to find out the causes of a problem before solutions can be introduced. An in-depth report must be prepared so that management is fully informed of the situation.

Students need the experience in preparing reports that require collection of data followed by an analysis of information gathered and conclusions based on the facts. A variety of techniques can be used to obtain the data; namely, interviews, surveys, or experiments.

This chapter will review the components of a formal report and illustrate a model of an informative report.

Steps in Writing a Report

The steps in writing a report follow:

1. Prepare an outline (a plan to help focus the purpose and structure of the report).

2. Gather data.

   Primary sources: surveys, telephone interviews, experiments

   Secondary sources: published materials, newspaper articles, government documents, online resources

3. Sort and analyze data.
4. Draft the report.
5. Proofread and edit the draft copy for the final text.
6. Use proper documentation (preferred style manuals are the MLA *Style Manual and The Chicago Manual of Style*). To ensure proper format refer to a style manual.

In the MLA document style, the source immediately follows the citation in the report.

Example: In 2003, worldwide online sales were about $80 billion (U.S. Bureau of Statistics, 10).

The complete reference appears in Works Cited.

The Chicago style uses footnotes or endnotes.

Example: . . . $80 billion.[1]

The complete reference appears in a footnote or in endnotes.

Parts of a Formal Report

There are different types of formal reports. Some reports, such as the sample in this chapter, are informative and use secondary sources to collect data. It is essential to obtain the latest data and relevant information before writing the report. There are many sources of secondary information. The library is an excellent source to examine newspapers, periodicals, government publications, and electronic databases. The Internet is another essential source to obtain current news, industry trends, and company-related data. Many companies offer helpful web sites.

Some business reports require research that uses primary sources; such as interviews, surveys, and experiments. This type of firsthand data lends support to analytical and scientific findings. Remember, one must always give credit to the source of data or a cited quotation. Plagiarism is a serious infraction in the publishing world. It can be easily avoided by using proper reference documentation as presented in style manuals.

Based on the type of report, sections may differ; however, most reports include the following categories:

- Title Page
- Table of Contents
- Text
  - Introduction
  - Body
  - Findings
  - Analysis
- Conclusions and Recommendations
- Endnotes, if applicable
- Appendix, if applicable
- Bibliography or Works Cited

Writing Pointers—Formal Report

- Report facts.
- Refrain from personal opinions.
- Document all quotations or statistics with footnotes, endnotes, or textnotes.
- Refer to an updated style manual to format references correctly.
- Place endnotes on a separate page at the end of the report. Key endnotes in the order in which the notation numbers appear in the text.
- Include side headings for clarity.
- Use visual aids, if necessary, to clarify information to reader.
- Prepare report free of errors in acceptable style.

Model Outline

TWENTY-FIRST CENTURY TRAVELING

I. Introduction
   A. Travel in the early part of the twentieth century
   B. Recent developments in travel

II. The Years Beyond 2000
   A. Reasons for explosion in travel industry
   B. Travel expenditures
   C. Predictions for leisure travel
      1. out-of-way places
      2. eco-travel
      3. health/wellness tourism

III. Impact of Technology
   A. Online travel market
   B. Use of Internet by airlines & hotels
   C. Discount rates available to consumers

IV. Redesign of Airline Terminals
   A. Business centers
   B. Children's play facilities
   C. Shopping and restaurants
   D. Transportation

V. Conclusions

Model Formal Informational Report

Title Page

TWENTY-FIRST CENTURY TRAVELING

Submitted to
American Travel Association
New York, NY 10021–3523

Submitted by
(Your name)

Current Date

Table of Contents

TABLE OF CONTENTS

Text

TWENTY-FIRST CENTURY TRAVELING

Introduction

Informational report

What a difference a century makes! Travel in the early part of the 1900s was often an arduous task. A cross-country trip on the railroad took over a week. The means of transportation from Europe to the United States was by steamship. As air travel expanded in the mid-1900s, business and leisure trips to Europe and the Far East became more accessible.

With the new millennium, the popularity of the Internet directly impacted the travel industry. Major changes were in store for both the traveler and the provider.

Side heading to focus attention

THE YEARS BEYOND 2000

The explosion in the travel industry has resulted from booming economies, easy access to global communications, and speed in reaching a destination. By the year 2005, it is estimated that travel expenditures in the United States will be about $594.3 billion.[1]

Secondary source

Endnote notation

Travel experts predict that out-of-the-way places such as Asia and Africa will be popular with the adventurous. Travelers who are interested in the environment will be attracted to eco-travel tours that visit natural, unspoiled areas.

Text provides information to support topic of report

The health/wellness tourism trend has been growing for some time. With a revived economy, more individuals are opting to combine a wellness holiday at a spa with other activities such as cooking classes, archeological study tours, and cultural activities.

IMPACT OF TECHNOLOGY

Facts explain side heading

The rapid growth of technology has had a dramatic influence on the travel industry. Through the Internet, people have access to travel information, which leads to a new world of sights and unknown places. In 2003, worldwide online travel sales were about $80 billion.[2] The Internet has become a major mode for making travel -

Endnote notation documents facts

2individuals can purchase airline tickets, reserve hotel rooms, rent cars, and book tours at discount rates. It has replaced some of the functions of the travel agent. Through the Internet, airlines and hotels have found that they have a direct distribution system to the consumer, simultaneously cutting costs and offering customers attractive package deals.

THE REDESIGN OF AIRLINE TERMINALS

Informative side heading

In the past, accommodations and services at airports were often unpleasant. To improve this situation, airline terminals are undergoing complete renovations. The new designs allow passengers to carry on business or personal activities in a pleasant, safe environment. Business centers equipped with data ports, computers, and fax machines can be found in the larger airports.

Presents information related to side heading

Airline terminals are becoming family-friendly by offering play facilities for children. At Chicago O'Hare, there is an interactive climbing experience created by the Children's Museum.

Example supports text information

Shopping is big business at airports. Brand name stores along with boutiques have opened up attractive places to shop at competitive in-town prices. For meals, passengers have choices from upscale restaurants to fast-food eateries.

Transportation to and within the airport is also undergoing major changes. Monorails connect downtown cities to airline terminals so those travelers can bypass traffic congestion.

CONCLUSION

The travel industry will continue to face challenges during the next decade. Companies that are innovative and sensitive to the needs of the present-day savvy traveler will be the leaders of the future.

Concluding statement of report

ENDNOTES

1. TIA Research Travel Forecast Summary, Travel Industry Association of America, Washington, DC. Available from <*http://www.tia.org/travel/forecasts.asp*.>

2. Travel Industry Worldwide, *eMarketeer*, November 2003.

Chicago Style reference notations

CASE STUDY FORMAL REPORT

In this chapter, you are required to prepare a *formal report*. Follow the "Writing Pointers" and guidelines given in this chapter. Use the model of a formal informative report in this chapter as a guide. Suggested topics you may use in developing this report follow. However, feel free to write a report on a topic of interest to you or with which you are familiar.

You will need to research the project to support your statements.

Suggestions

- Choose a locale of interest to you. This may be a city, state, country, or historic or vacation site. Research the topic and gather the data to be used in writing an informative report.
- Compare two types of cruise lines that would include an analysis of the company, number of passengers served over a period of time, facilities, types of cruise ships, services on board the ship, types of tours, and prices. Some of the data could best be displayed in tables.
- Write an historical report on a mode of travel (automobile, railroad, air).

ASSESSMENT: CHAPTER 10

English Power

Directions: *In each sentence below, select the proper form of the pronouns* who *or* whom *and write it in the space provided.*

____________ 1. The cruise director (who, whom) I contacted planned our itinerary.

____________ 2. The major airlines are offering discounts to travelers (who, whom) will book by October 1.

____________ 3. To (who, whom) should this voucher be sent?

____________ 4. The South American tour of the rain forests is attractive to those (who, whom) are interested in the environment.

____________ 5. The agent (who, whom) handled our reservations was polite and helpful.

____________ 6. The advertisement on discounted airfares to the West Coast appealed to those individuals (who, whom) were planning a vacation in California.

____________ 7. The hotel is situated on five acres and is perfect for people (who, whom) like to take daily walks around a lake.

____________ 8. European airpasses are available for purchase only outside Europe and to those individuals (who, whom) are not European.

____________ 9. We wish to hire prospective employees (who, whom) we believe we can train.

____________ 10. (Who, Whom) did the World-Wide Travel Association elect as its new president?

Editing Power

Directions: *Edit, format, and proofread the following unarranged draft form letter. Use proofreaders' symbols to mark the errors. Use "Regional Director:" as the salutation, supply missing parts and paragraphs, and key the letter correctly.*

```
Every year thousands of visitors whom spend there
vacation in the south visit our beaches lakes and historic
districts. As we observe National Tourism Week May 2-8
let us keep tow factors in mind. First our states are
known throughout the country for the4ir hospitality.
This reputation helps keep tourism one of the largest
industries in the South. Second each resident is effected
by visitors who have an impact on the souths economy.
Revenue from tourism and travel help support jobs and
lowers taxes. We are planning to issue a special edition
of our travelers guide during National Tourism Week. it
will highlight some of the Souths interesting sites. We
invite you to submit a 100-word article describing
a geographic area or attraction which would interest
tourists.^You may want to include a photograph which
capture the eyes of the reader. When preparing your
material keep in mind your target population. We look
forward to receiving your information by March 1 so
```

```
forward to receiving your information by March 1 so
that we can meet our printing dead line. A popular
feature with tourists has always been a description of a
scenic automobile route for those whom enjoy driving.
```

Writing Power

Directions: *After reading the case below, draft a letter to* Ms. Ellen Dow, 15 Fieldstone Drive, Butte, MT 59701-3634. *Ms. Dow, a prospective customer, requested information on your firm's prepackaged tours for the coming summer. Compose a cover letter to include the following unique features of these tours:*

- *escorted tours with knowledgeable guides*
- *scheduled air flights*
- *sightseeing "off the beaten track"*

You may include other facts that are appealing to a traveler. Enclose a descriptive brochure with the letter.

"If your failing to plan, you are planning to fail."

Tariq Siddigue

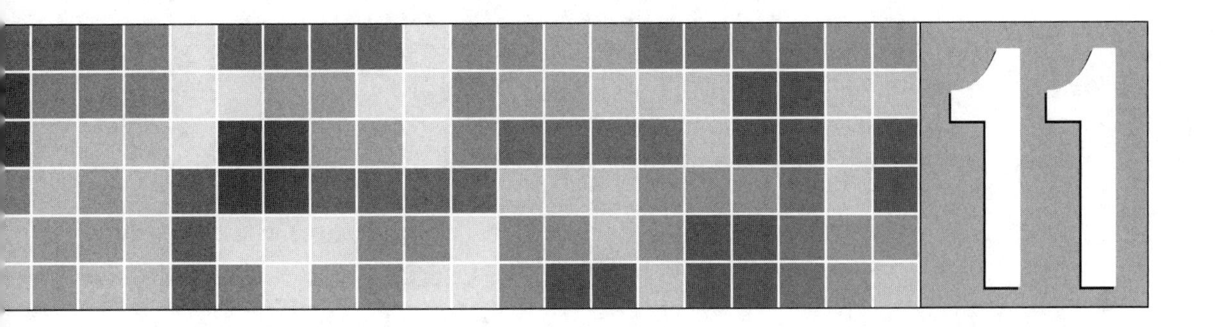

Governing Quotations

WHAT YOU WILL LEARN IN THIS CHAPTER

- To apply quotation marks correctly
- To apply correct punctuation when using quotation marks
- To develop a business vocabulary in **GOVERNMENT**
- To proofread correspondence and reports
- To compose community public relations letters
- To compose an informational letter
- To compose a citizen's complaint letter to a government official
- To compose a personal complaint letter to a government official
- To compose a reply to a citizen's request
- To write an e-mail memo
- To compose a news release
- To compose minutes of meetings

Corporate Snapshot

Best Places to Live

THROUGHOUT the United States, many communities have been recognized as best places to live. What criteria determine a desirable town or city?

Among the factors to consider are job and income growth, educated workforce, affordable housing, good educational facilities, cultural and sports activities, low crime rate, community involvement, and a vision for the future. The following communities are samples that have been cited as great places to live.

Portland, Oregon: The citizens are proud of their city's transformation from a timber town to a high-tech hub. Keen city planning has given rise to an urban metropolis incorporating natural environment, transportation, parks, and neighborhoods. Within an hour of the city, its residents can be on a mountain, at the beach, in the desert, or in the natural forest.

Anchorage, Alaska: As a tent city in 1957, Anchorage exploded in the 1970s with the construction of the Trans-Alaskan Pipeline. No longer dependent on the oil industry, Anchorage has a diversified economy and a steady increase in population. Tourism has become the major industry for this city known as the "crossroads of the world." Its residents on one hand can enjoy Anchorage's cosmopolitan lifestyle; on the other, they take pleasure in the incredible natural wonders of the surrounding areas.

Sugarland Run, Virginia: This small town with a population of under 30,000 was listed as a top-ranked place to live. Sugarland Run is horse country close to the Potomac River and Washington, DC. Newcomers are attracted to the area. It has good employment opportunities, fine schools, and affordable housing options. At the same time, the population is within a short drive of the cultural life of the capital.

Sources: *http://money.cnn.com/best/bplive/portland.html*
http://money.cnn.com/2003/12/08/pf/bplive 03_east/index.htm
Frontier Airlines Magazine, May-June, 2004

ACTIVITY A LANGUAGE ARTS RECALL

Quotation Marks

Quotation marks indicate a speaker's exact words. Quotation marks are used at the beginning and end of the quoted phrases or statements. This is called *direct speech*.

Note: Quotation marks are not used for a paraphrase or a restatement of a quotation, called *reported speech*. Paraphrases are often introduced by the word *that*.

Rule 11.1: Use quotation marks around a person's exact words.

Example: Mr. Eric Johnson stated, ***"Satellite discs will be a hot issue at the next meeting."*** (direct speech)

Mr. Johnson said that a discussion on satellite discs will be on the agenda next month. (paraphrase/reported speech)

Note: A comma usually precedes a quotation. The first word of the quotation is capitalized.

Rule 11.2: Use quotation marks around each segment of the quotation when there is an interruption. Note that the first word of the second segment of the quotation begins with a lowercase letter unless it is a proper noun. Set off the interrupting expression with commas.

Example: ***"We need to call a meeting,"*** the town supervisor said, ***"to discuss zoning procedures for commercial development."***

Rule 11.3: Use *single quotation marks* for words quoted within a quotation.

Example: The guest speaker asked, "Has anyone read the pamphlet ***'Tips for the Consumer'***?"

Rule 11.4: Use quotation marks around the titles of complete but unpublished works, such as reports, manuscripts, programs, and booklets. However, underline or italicize the titles of complete published works, such as books, magazines, and newspapers.

Example: You will want to read the report "*The New Direction of Government Spending*" that appeared in *The Daily Record.*

Rule 11.5: Use quotation marks around titles that represent parts of published works, such as chapters in a book, lessons, and articles.

Example: The chapter "*Community Volunteerism*" that is found in the best seller *Government Update* is very informative.

Punctuation at the End of Quotation Marks

Rule 11.6: When a sentence ends in a period or comma, the period or comma is always placed inside a quotation mark.

Example: Mary Kelly stated, "A major construction project on our main highway will begin at the end of the year."

Rule 11.7: Colons and semicolons are always placed outside the quotation mark.

Example: The flyer read, "The conference fee is $59"; however, the special group rate was not mentioned.

Rule 11.8: Question marks and exclamation points are placed either inside or outside the quotation marks depending upon the meaning of the sentence.

Place a question mark or exclamation point inside the quotation marks when it applies to the sentence that is within the quotation marks.

Example: The professor asked, "Is there a difference in earnings between union and nonunion workers?"

When the question mark or exclamation point applies to the entire sentence, it goes outside the quotation mark.

Example: Did Ramon say, "The unemployment rates in the various sections of the country are decreasing"?

Check for Understanding

Directions: Change each of the following sentences from reported speech to direct speech.

Example: The town clerk announced that she will run for the office of town supervisor.

Rewrite: The town clerk announced, "I will run for the office of town supervisor."

1. The mayor stated that there must be a reduction in municipal spending.

   Rewrite: ____

   ____

2. Our county legislator asked about ways in which government spending can be reduced.

   Rewrite: ____

   ____

3. The congresswoman commented that legislation will be enacted to assist single parents.

   Rewrite: ____

   ____

4. The guest speaker wanted to know if anyone read the pamphlet on government programs.

   Rewrite: ____

   ____

5. The repaving of Main Street will begin on Monday, according to a statement by a spokesperson of the traffic department.

   Rewrite: ____

   ____

(Answers are given below.)

Answers to "Check for Understanding"

1. The mayor stated, "There must be a reduction in municipal spending."
2. Our county legislator asked, "How can government spending be reduced?"
3. The congresswoman commented, "Legislation will be enacted to assist single parents."
4. The guest speaker asked, "Did anyone read the pamphlets on government programs?"
5. "The repaving of Main Street will begin on Monday," said a spokesperson for the traffic department.

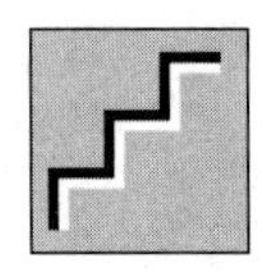

Building English Power: Quotation Marks

Directions: Insert quotation marks and other punctuation where needed.

1. The town supervisor stated If you have any suggestions on ways in which we can improve the town please contact my office. (11.1, 11.6, 6.1)
2. Property taxes will increase by 10 percent stated the controller unless we reduce spending. (11.2, 11.6)
3. Are you interested in the newly formed civic association asked my neighbor. (11.1, 11.8)
4. The judge suggested that a community service program could be an alternative means of punishment for minor offenses. (11.1)
5. The notice read as follows A public hearing to discuss housing will be held on April 15 2005 at Town Hall. (11.1, 10.1, 7.8)
6. You an active civic leader are invited to attend the meeting on Thursday stated the council president. (11.1, 11.6, 7.5, 1.4)
7. I asked Have you seen the article titled federal budget that appeared in the Tribune Herald last week (11.1, 11.3, 11.4, 11.5, 1.6)
8. Did the community leader say We need to have greater recreation facilities for our youth (11.8)
9. Do you plan to attend the program called multicultural interaction that is being sponsored by the Community Development Department (11.4, 11.1)
10. As you know commented the congresswoman we are introducing legislation to give financial assistance to college students. (11.2, 11.6, 6.1)
11. My friend asked Have you read the article The Role of Government in the 1990s (11.1, 11.3, 11.5, 11.8)
12. With the changes in traffic patterns announced the legislator we need to revise our noise ordinance. (11.2, 11.6)
13. Is it possible that the trial will be postponed asked the lawyer. (11.8)
14. In honoring the town resident Jim O'Connor the mayor said Jims dedication to his neighbors has helped our quality of life. (11.1, 11.6, 7.6, 6.3, 9.1)
15. The civic association member asked Why were the beautification plans for Main Street postponed until the spring (11.1, 11.8)

ACTIVITY B BUILDING WORD POWER

The words below are commonly used in the field of government. From this list, select the word or phrase that correctly completes each statement. Refer to the Personal Dictionary, if necessary.

alternative	legislation
assessment	legislator
budget	multicultural
bulk	municipal
commissioner	ordinance
diminished	stalled
fiscal	town council

_______________ 1. The ________ year for our town ends in July; therefore, all monies allocated must be spent before that date.

_______________ 2. The town is sponsoring a(n) ________ fair featuring ethnic foods and entertainment that represent the many different nationalities in our community.

_______________ 3. The main function of Congress is to vote on new ________; that is, to pass proposals that become laws.

_______________ 4. Legislation was ________ in the Senate because the members could not reach a majority vote; therefore, the law did not pass.

_______________ 5. The rate of inflation has ________ greatly in the past decade. It has declined to its lowest point.

_______________ 6. Both individuals and government need to have a(n) ________ so that they can plan for expenditures based on their income.

_______________ 7. The first woman in our community elected as ________ was assigned to chair the Planning Board.

_______________ 8. The ________ is made up of legislators who are elected to govern the community.

_______________ 9. In this year's budget, the city must increase monies for ________ services, such as fire and police.

_______________ 10. The mayor offered a(n) ________ plan to raise taxes, which differed from the town council's recommendation.

_______________ 11. The new recycling program requires that newspapers be packaged in ________ rather than disposed of individually.

_______________ 12. The town board passed a new ________ that prohibits loud noise after 11 p.m.

_______________ 13. The traffic ________, a public official, is responsible to the mayor for the progress of the bridge construction.

_______________ 14. The official value, or ________, of your house is based on the current market value.

On the Alert

- residents (plural noun)—people who live in a place

  Example: The town ***residents*** will vote on the school budget.

- residence (noun)—place to live

  Example: The governor of New Jersey has an official ***residence*** in Trenton.

- One-thought modifiers (compound adjectives) are hyphenated before the noun.

  Example: well-known public figures

- "Cannot" is always written as one word.

- time of day—When written in lowercase letters, the abbreviations "a.m." and "p.m." are separated by periods with no spacing within.

  Example: 10 a.m.

- Seasons are generally not capitalized unless they are personified.

  Examples: 1. The annual convention will be held in the ***spring.***

  2. Each October, ***Autumn*** spreads its leaves across the earth, draping it in color.

ACTIVITY C INDUSTRY PERSPECTIVES: FAMOUS QUOTATIONS REFLECTING AMERICAN ATTITUDES

Directions: Using appropriate revision marks, proofread and key the report below. The title is FAMOUS QUOTATIONS REFLECTING AMERICAN ATTITUDES.

Rule Reference

6.3 Over the years there have been many changes in our

7.1 society—politically socially and economically—which have come about through changing attitudes and beliefs. The following quotations from well known public figures reflect their thinking during the decades from the 1940s to the present.

THE DECADE OF THE '40S

Franklin D. Roosevelt, in the 1940s, stated in one of

11.1 his annual messages to Congress We have come to a clear realization of the fact that true individual freedom can

11.1 not exist without economic security and independence.

THE DECADE OF THE '60S

6.3 In the 1960s Martin Luther King's famous speech at

11.4 the Lincoln Memorial, I Have a Dream, gave the following

10.1, 11.1 message I have a dream that one day this nation will rise up and live out the true meaning of its creed, 'We

11.1 hold these truths to be self evident that all men are created equal'

THE DECADE OF THE '80S

6.3, 1.7 During the 1980s americans were concerned with the

9.1, 11.4 countrys ability to meet global competition. Business

11.1, 9.1 Week reported The nations ability to compete is threatened by inadequate investment in our most

11.6 important resource: people".

THE DECADE OF THE '90S

During the 1990s, the country became concerned about protecting its environment and natural resources. Helen

1.4 hayes captured the feelings of those Americans who love

11.1 nature in one of her poetic statements. It read All through the long winter I dream of my garden. On the first warm day of spring, I dig my fingers deep into the

11.6 soft earth. I can feel its energy, and my spirits soar

THE NEW MILLENIUM

9.1 The new millenium brought to the forefront Americas sensitivity to the cause of diversity and women. U.S. Supreme Court Justice Ruthe Bader Ginsberg reflected

11.1 this attitude when she said I expect to see the day when women serve the cause of Justice in numbers fully reflective of their talent. That day, when bias,

11.1 conscious or unconscious, is no longer part of the scene, has already dawned in some places.

Did you find these errors?
11 quotation marks
9 commas
2 hyphens
1 spelling/typographical
1 colon
2 capitalization
3 possessives
1 underscore or italicize
3 other punctuation

ACTIVITY D EDITING POWER

Task A: Edit, format, and proofread the following unarranged memorandum, then key it correctly. Remember to insert paragraphs.

Errors
- quotation marks
- commas
- capitalization
- colon
- word usage
- dates

TO: Civic Association Leaders, FROM: Paul Turner, SUBJECT: BUDGET HEARINGS. You are cordially invited to participate in the 20xx budget process. There will be two public meetings to discuss the budget proposal. The first meeting will be held on thursday October 15 20xx at 7 p.m. at City hall. The following departments will make their presentations Department of public works Parks and Recreation, and Planning and Zoning. The second meeting which will be held on november 1 at 7 pm will include the following departments police fire and purchasing. At these meetings the public will have an opportunity to ask questions on any item of the budget. In the passed people frequently have asked How come we were never informed about local issues. I firmly believe that local residents must be knowledgeable about there community. As I have stated previously public involvement is vital to effective functioning of government. I look forward to working with you as we begin to prepare for the next fiscal year.

Task B: Edit, format, and proofread the following handwritten form letter for all errors; key it correctly. Remember to insert a salutation and closing.

Many of the Greenfield residents have asked about my future in government now that my fifth term on the Town Council has ended. Over the passed few years I have watched our wonderful community slide backwards. Our taxes have doubled in five years yet our community services has diminished drastically. Greenfield has lost many of its experienced and competent employees. They felt that there voice was not being heard in the decision-making process. Important community projects that can increase our tax base has been stalled in committee. Recently, a local real estate agent commented Greenfield's property values are declining. Private citizens are heard asking Where is our leadership? During the past few months I have talked with many of you who share my vision of getting Greenfield back on track. Therefore I have decided to seek the office of Town Supervisor. With your help, I am confident that we once again can make Greenfield a great place to live. Please feel free to call me at my office (992–6000).

ACTIVITY E WRITING PRACTICUM

Prepositions

A preposition is a word that connects a noun or pronoun to another word in a sentence. Below are some examples of commonly used prepositions.

about	down	over
above	during	regarding
after	except	through
around	for	to
at	from	toward
before	in	under
behind	into	until
below	of	up
beside	off	with
by	on	without

among (used with three or more persons or things)
between (used with two persons or things)

A **prepositional phrase** consists of the preposition and its object.

Examples: The committee discussed the budget ***during the seminar.***
The city offices reopened ***after the storm.***

EXERCISE 1 USING PREPOSITIONS CORRECTLY

Complete each of the following sentences by selecting the correct preposition from within the parentheses. Check the meaning of unfamiliar words in the dictionary. Write the word in the space provided.

__________ 1. The Federal Government consists (of, in) three branches: executive, legislative, and judiciary.

__________ 2. The municipal judge noticed a discrepancy (among, between) the testimony given by the police officer and the crossing guard.

__________ 3. Rural counties deal (with, in) problems different from those of urban governments.

__________ 4. The minority does not have to agree (to, over) the majority decision.

__________ 5. The small downtown businesses asked their state representatives for help to avoid relocating (off, from) their present sites.

__________ 6. A celebration in the state capitol will be held later (in, on) the week.

__________ 7. Two-thirds of the voting members must agree (to, for) the proposed amendment.

__________ 8. Senior citizens may apply for a reduced tax assessment (on, from) their property.

__________ 9. The market value of homes like yours is based on recent sales of similar properties and also is used as a guide (for, of) new assessments.

__________ 10. The special committee argued (about, with) the terms of the contract.

__________ 11. The conservative party differed (about, from) the way community funds should be used.

____________ 12. The members of the city council could not agree (between, among) themselves about the legal wording of the ordinance.

____________ 13. The Zoning Board's report was accompanied (with, by) a video presentation.

____________ 14. The merchants were angry (at, over) the Traffic Commissioner for rerouting the downtown traffic.

____________ 15. The proposal for a community center was introduced (to, during) the meeting.

Puzzling Words

Some words are troublesome to a writer. To help you avoid making some common errors in word usage, study the following words and definitions; then complete the exercise after these words.

a lot	(adjective) a great number (slang; do not use)
allot	(verb) to assign
among	(preposition) in a group of three or more
between	(preposition) each with one other (only two)
anyone	(pronoun) a person
any one	(pronoun) an individual person or thing
can	(verb) used to indicate ability or power
may	(verb) used to indicate permission or possibility
everyone	(pronoun) everybody
every one	(pronoun) each person in a group
	(Note: Spell the pronouns *anyone* and *everyone* as two words when they are followed by an *of* phrase and mean one of a number of things.)
fewer	(adjective) refers to amount and used with countable nouns
less	(adjective) refers to degree or amount and used with noncountable nouns
lay	(verb) to place or put something down (always takes an object)
laid	(verb) past tense of lay
lie	(verb) to rest or recline

lose	(verb) to be unable to find; to mislay
loose	(adjective) not fastened; not close together
irregardless	*not acceptable usage*
regardless	(adjective) showing no regard
set	(verb) to place something somewhere
sit	(verb) to be seated
some time	(adjective–noun) period of time
sometime	(adverb) an indefinite time in the future
their	(pronoun) possessive form of "they"
there	(adverb) in a place; used to introduce a clause
they're	contraction for "they are"

EXERCISE 2 PUZZLING WORDS

Select the word or phrase that correctly completes each statement, then write it in the space provided.

_______________ 1. We have (fewer, less) people willing to serve on the election committee.

____

_______________ 2. The candidate has been given (fewer, less) time to address the TV audience.

____

_______________ 3. You (may, can) attend the seminar if you are a member of the organization and have made advance reservations.

____

_______________ 4. Only people who have the financial resources (can, may) afford to be candidates for government offices.

____

_______________ 5. The speaker (set, sit) the microphone on the lectern.

____

______________ 6. The prospective jurors will (set, sit) in the main waiting room.

____

______________ 7. You may want to get insurance in case you (lose, loose) your luggage on the trip.

____

______________ 8. The (lose, loose) door handle caused the car rattle.

____

______________ 9. The club was to (a lot, allot) circus tickets to the underprivileged children.

____

______________ 10. The suspension of the two employees was caused by a disagreement (between, among) them.

____

______________ 11. The discussion on a fare increase was held (between, among) the community residents.

____

______________ 12. (Some time, Sometime) next year, I plan to take a trip to Alaska.

____

______________ 13. The travel agent spent (some time, sometime) with the client in planning a vacation trip to the West Coast.

____

______________ 14. (They're, Their, There) planning to publish a travel newsletter in the spring.

____

______________ 15. The couple planned (they're, their, there) vacation by surfing the Internet.

____

______ 16. (Any one, Anyone) who resides in this community is eligible for a reduced parking rate.

______

______ 17. (Any one, Anyone) of our state employees is eligible for a pension after 20 years of service.

______

______ 18. A token gift was given to (every one, everyone) in attendance at the festival.

______

______ 19. (Everyone, Every one) of the candidates proposed open space legislation to maintain park land.

______

______ 20. (Regardless, Irregardless) of public opinion against a property tax increase, it was approved by the assembly vote.

______

______ 21. The librarian (lay, lie) the historical document on the table for our review.

______

______ 22. After the strenuous exercise class, my friend decided to (lay, lie) down before going to dinner.

______

Writing Pointers

Community Public Relations Letter

- Stress importance of the meeting/activity.
- Appeal to sense of community responsibility.
- Impress on readers that they can make a difference.
- End with statement that encourages reader to become involved.
- Maintain support of constituents by keeping them informed of legislative actions.

Model Information Public Relations Letter to County Constituents

Date

MS NORA BRODY
177 EAST MAIN STREET
IRVINGTON NY 10524-3213

Dear Ms. Brody:

Demonstrates interest in community welfare

The recent survey completed by the residents of Greenburgh County revealed broad support for the preservation of open space in our communities. A clear mandate was expressed to end overdevelopment of shopping malls and large office parks in the towns.

Explains action to be taken

At the county and state levels, I intend to use the findings of the survey to enact immediate legislation to prevent the depletion of open space. Local government needs to reexamine its policies and practices with regard to zoning for building, traffic patterns, and quality of life. At the next legislative meeting, I will request state funds to secure parcels of land that will be used for parkland. This initiative will ensure open spaces to protect the environment for future generations.

Appeals to sense of community cooperation

Offers to provide further assistance

As we work together, I am confident that we shall maintain the quality of life that has made Greenburgh County a desirable place to live. If you have any further concerns, do not hesitate to contact my local office at 479-3000.

Sincerely yours,

Richard L. Pearce
Assemblyman

rlc

ACTIVITY F WRITING POWER: COMPOSING CORRESPONDENCE

CASE STUDY 1: INFORMATIONAL LETTER TO COMMUNITY

After reading the following case, respond to the questions in "Organize Your Ideas" before beginning to compose this letter. Then draft the letter.

Case: A cover form letter from Blanche Mitchell, town clerk, is to be mailed to the community leaders informing them that the revised color-coded Sanitation Schedule is enclosed. This schedule includes the following information:

- Pickup days for refuse
- Pickup days for recycling of bottles and plastics
- Listing of telephone numbers of town offices

Sanitation schedules are available in bulk at Town Hall to civic associations for distribution in their neighborhoods. In the letter, ask for the community's cooperation in maintaining a clean, healthy environment.

Organize Your Ideas

1. What is the subject of this letter?

____

2. What would you include in the opening sentence?

____

____

3. What specific details would you include in the body of the letter?

____

____

____

4. What would you say in your closing statement?

____

____

CASE STUDY 2: PUBLIC RELATIONS INFORMATIONAL LETTER TO TOWN RESIDENTS

After reading the following case, respond to the questions in "Organize Your Ideas," then draft the letter.

Case: Howard Wallace, town supervisor, asked that you send out a form letter to the residents of Smithport informing them that there will be a public hearing on the needs of senior citizens. The letter will include the following information:

Agenda topics:

- updating present programs
- future services
- assessment of recreational facilities

Location: Community Center

Date: Monday, May 5

Time: 7 p.m.

Transportation to the Community Center: Call 592–2916

Organize Your Ideas

1. What is the main topic of the letter?

____

2. What would you include in the opening sentence?

____

____

3. What specific details would you include in the body of the letter?

____

____

____

4. What would you say in your closing statement?

____

____

CASE STUDY 3: INFORMATIONAL LETTER

Read the case below, then draft the letter.

Case: The town supervisor, Paul Murphy, has asked you to write a cover letter that will be attached to the proposed annual town budget. In the letter, explain that the 2.5 percent increase in taxes is the lowest in the past five years. Also, mention that there has been an increase in police and recreational services. Invite the readers to attend the town board meeting on the budget to be held on the first Monday of next month. (Insert the correct date of the meeting.) Close the letter by asking the town residents to support the proposed budget. Address the letter to the "Town Residents."

CASE STUDY 4: PUBLIC RELATIONS LETTER TO CONCERNED RESIDENTS

Read the case below, then draft the letter.

Case: In the city of Oakridge, the residents have voiced their concern about the lack of after-school educational and recreational services for school-age children. They fear this will result in an increase in the social problems that will arise in the community. To prevent this from occurring and to keep the community safe, one proposal suggests offering reading courses, gymnastics, and team sports at the local school from 3 to 5 p.m. weekdays.

Pauline Slater, chairperson of the City Council, is writing to the residents inviting them to an informational meeting on Tuesday, April 10, (current year), at City Hall to discuss the issue, consider possible solutions, and raise funds for such a program. The residents are being asked for their input in designing a plan. Draft a letter from Ms. Slater.

CASE STUDY 5: RESPONSE TO CITIZENS' REQUEST

Read the case study; then compose the letter.

Case: Anna Lorenzo, town supervisor, is responding to a request from the residents on Cooper Street. They want a traffic light to be installed at the intersection of Cooper Street and Brook Avenue. Numerous accidents have occurred at the site in the past year. Motorists do not adhere to the stop signs and race through the intersection. The latest casualty occurred a week ago when a 12-year-old youngster was severely hurt and remains in serious condition.

Ms. Lorenzo acknowledges the seriousness of the situation. However, it is a more complex problem. Traffic lights are under the jurisdiction of the state, and a study will need to be conducted by traffic engineers. Based on their findings, a decision will be made.

Ms. Lorenzo will petition the state to install a traffic light. In the meantime, she will request an additional police patrol be assigned to the area.

Prepare a form letter for Ms. Lorenzo. Use the salutation Dear Resident:

CASE STUDY 6: E-MAIL MEMORANDUM

Read the case study; then compose the e-mail memo.

Case: Sophia Cruz, town supervisor of Pomona, is writing this e-mail to the Town Board members informing them of the meeting she held with the merchants on Main Street with regard to the revitalization plan for the downtown area. Ms. Cruz received an excellent response from the merchants with the first phase of the plan that will involve the resurfacing of the sidewalks with bricks and planting of trees. The group also agreed to redo the facades of their stores so that there would be a uniform hometown feeling. The town will pay one-half of the cost. This phase will start next summer after a design is approved. The next meeting will be held at the start of the resurfacing.

Prepare an e-mail memo. The list of Town Board members who are to receive the memo will be taken from the address book. Leave the "To" line blank.

ACTIVITY G REPORT WRITING POWER: NEWS RELEASE

A news release is a document sent to the media (newspaper, magazine, radio, and TV) to convey timely news that is of interest to readers. The content of a news release may include an introduction of new products or services, organizational or personnel changes, staff accomplishments, and firm's financial status.

A news release is an excellent way to build good public relations.

Writing Pointers for News Releases

- Include a headline to identify story.
- Give most important message immediately.
- Be brief, specific, and to the point.
- Avoid any marketing or advertising pitch.
- Double space document.
- Indicate the time when news can be released (For Immediate Release).
- Include contact person, name, and telephone number.
- End news release with # # # centered on line.

Model News Release

ALBERTVILLE TOWN BOARD
20 Fairview Road
Albertville, MD 21400-2100

Special instructions

For Immediate Release

Contact person

From: Emanuel Delgado
Director, Public Relations
(410) 832-6021

NEWS RELEASE

Title highlights topic

PURCHASE OF THE GREENWALL ESTATE

Dateline

States most important fact

Specific information

ALBERTVILLE, MD, December 10, 20xx. James Peterson, Town Supervisor, announced last night at the Town Board meeting that an agreement had been reached to purchase the Greenwall Estate for $10 million. The purchase of the property is a victory for the town and the environmentalists who feared that the beautiful park-like site would be sold to developers. The purchase was made possible through private fund-raising.

Quotation projects a personal touch

Positive closing

In keeping with the open-space policy of the town, Supervisor Peterson said, "We are committed to maintaining the quality of life for all families of Albertville. The purchase of the Greenwall Estate will assure our citizens the enjoyment of a new park for many years."

Symbol indicates end of release

# # #

CASE STUDY 1: NEWS RELEASE

After reading the case, respond to the questions in "Organize Your Ideas," then compose the news release.

Case: As the assistant to the Commissioner of Environmental Services, compose a news release announcing that the town of Hartsview, New York, will adopt a recycling program for the community. Explain that such a program is essential to maintain the quality of life by reducing pollution.

The pickup schedule will be every first and third Tuesday of the month. Special containers will be distributed to each household. Include a phone number for further information.

This announcement is being made for immediate release by Larry McGuire, Commissioner of Environmental Services.

Organize Your Ideas

To help you write an effective news release, respond to the following questions before composing it.

1. What title would you give your news release?

   ____

2. What information must appear in the news release?

   ____

3. Which of the following approaches would give a human interest appeal to the news release?
   a. Quote from the Commissioner of Environmental Services
   b. Lecture readers on the importance of recycling
   c. Present news on recycling from neighborhood communities

   ____

4. What other statement would you include in your closing paragraph?
   a. Citizens' reaction to recycling plan
   b. Appeal to community for cooperation
   c. End news release after explaining project

   ____

Report Writing Power: Minutes of Meeting

Minutes of meetings are the written record of the proceedings of a meeting. Accurate recording is essential because minutes serve as a permanent record of decisions that were made and actions taken. Individuals who did not attend the meeting use the minutes as a resource to become familiar with the business that occurred.

Formats for producing minutes vary. However, they are uniform in requiring all topics covered to be in chronological order and to include exact quotes from participants when necessary. Read the "Writing Pointers" to learn the fine points in developing minutes of a meeting.

Writing Pointers

Minutes of meetings

- Indicate time and place of meeting.
- List the names of those individuals who were present and absent.
- Mention whether previous minutes were read, approved, amended, or not approved.
- Record important facts discussed at the meeting. The body of the minutes should include all old and new business.
- Record verbatim resolutions and motions. Include the names of those people who introduced and seconded motions.
- Indicate the date and time of the next meeting in the closing paragraph as well as the time of adjournment.
- End with words "Submitted by" or "Respectfully submitted."
- Include the name and signature of the person who writes the minutes.

Model Minutes of Meeting

ALBERTVILLE TOWN BOARD

MINUTES OF THE MEETING
January 4, 20xx
Town Hall, 8 p.m.

Individuals in attendance

PRESENT: Town Supervisor James Peterson, Councilman Ray Adler, Councilman Edward Barnes, Councilwoman Sandra Parker, and Councilwoman Lisa Choy.

ABSENT: Town Justice Deborah Coles

Approval of minutes

The minutes of the previous meeting held on December 4, 20xx, were approved as distributed.

OLD BUSINESS

Topics previously discussed

Town Supervisor James Peterson reported on the Open-Space Committee meeting held last week. The committee voted to study the possibility of purchasing the ten-acre vacant lot on Eastern Boulevard to be developed into a "pocket park."

Police Commissioner Reynolds discussed the duties of the newly appointed street patrol officer. This appointment was made possible as a result of a federal grant to increase police protection in communities. One of the most important duties for this police officer is to establish a rapport between the police and the merchants.

Motion made and approved

Motion was made by Ray Adler and seconded by Lisa Choy that E. R. Transport Co. be awarded a one-year contract for $35,000 to transport senior citizens to recreational and shopping centers. Motion carried.

NEW BUSINESS

Topics presented

Refurbishing the children's playground was introduced by parents of the community. The Albertville Town Board tabled the issue.

A public hearing would be held at the next meeting to discuss the adoption of new fire codes for public buildings.

ADJOURNMENT

The meeting was adjourned at 10 p.m. The next meeting is scheduled for February 15, 20xx.

Respectfully submitted,

Shirley Schuller

Shirley Schuller, Secretary

CASE STUDY 2: MINUTES OF MEETING

After reading the case, respond to the questions in "Organizing Your Ideas," then compose the minutes of the meeting.

Case: You were given an assignment to cover the monthly meeting of the Student Government to be held on January 15, 20xx, at 1 p.m. at your school. Prepare the minutes of this meeting.

All the officers were present: Vicki Forsythe, Daniel Rhodes, Gloria Rockport, and Chris DeAngelis. Minutes of the previous meeting were approved.

Leslie Dunn reports that 50 students volunteered to work on the Local Disaster Committee to raise funds for the victims of the recent floods in the community. The committee plans to collect money from merchants, shoppers at the mall, and the school. Posters and flyers will be distributed throughout the neighborhood. The group will start its drive next week.

Robert Jones reports that he is representing the students at the Faculty Senate. One of the topics under discussion is raising academic standards. Subcommittees will be formed to explore various aspects of academic achievement. Each subcommittee will have a student representative.

A new topic on the agenda is Brian Scott's request that the Student Government look into the possibility of extending the hours of the

Media Center. It would also help students who work to keep the Center open on Sundays.

The Political Science Club proposes that Student Government should plan a debate and invite the local candidates running for office. This will be discussed at the next meeting.

The meeting was adjourned at 2:30 p.m. The next meeting will be held the following week at the same time.

Organize Your Ideas

To help you write effective minutes for the meeting, respond to the questions below.

1. What are the subheadings to use when writing the minutes of the meeting?
2. What are the major topics to be reported under Old Business?
3. What are the major topics to be reported under New Business?
4. What should you write in the closing statement?

ASSESSMENT: CHAPTER 11

English Power

Directions: *In the following sentences, insert quotation marks and other punctuation where needed, then key each sentence.*

1. When will they install the new traffic light on Groves Road asked the resident.
2. The public hearing about the revised town building code has been set for June 1 stated the council clerk.
3. The police chief said Interaction between community and police will promote better understanding and enhance sensitivity.
4. The article Revitalizing Local Neighborhoods focused on increasing community involvement.
5. An announcement was made that a federal grant for community policing has been received for the Mayfield area of town.
6. With the approaching summer months commented the county executive we need to implement a youth job program.
7. It is our understanding said the commissioner that the panel will interview prospective applicants for the position of Housing Authority supervisor.
8. The property owner asked the city council Why was there a 10 percent increase in real estate taxes.
9. The author stated I believe in a grassroots approach to government as discussed in my article Back to the Basics in Politics.
10. The judge said It is the civic duty of all citizens to serve on a jury.

Editing Power

Directions: *Edit, format, and proofread the following form letter. Use "Dear Residents:" as the salutation, supply missing parts and paragraphs, and key it correctly.*

The greenfield Environmental Council is sponsoring an Earth Day celebration for our community on Saturday april 20. The festivities will begin at 10 am with a brief ceremony, which will take place at the Town Plaza, featuring our high school band. After that, we will pick up litter along the main road and plant flowers in the Town Plaza. Earth Day remind us of our civic responsibility to stop pollution and preserve our natural resources air water and soil. Evans Robinson the well known environmentalist summarized our objective when he said Be involved and reclaim our environment. If you wish to participate please call my office. See you on April 10!

Writing Power

Directions: *Compose a letter to the town residents. Invite them to a public hearing on the beautification of Cutter Mill Park, the town park on Willow Lane. Tell the residents that funds have been allocated by the Planning Board for this project. Proposals on topics such as new landscaping, replanting trees and flowers, and building a children's playground will be discussed. Date of the meeting is Tuesday, June 5; place is Town Hall; time is 8 p.m.*

There is no "I" in TEAMWORK.

Heartquiotes.net

12

Help Wanted: Applying Hyphens and Numbers Correctly

WHAT YOU WILL LEARN IN THIS CHAPTER

- To apply hyphens correctly
- To write numbers correctly
- To develop a business vocabulary in **HUMAN RESOURCES**
- To proofread and edit correspondence and reports
- To compose correspondence related to the job search
- To prepare a resume

Corporate Snapshot

Kelly Services, Inc., is an international leader in providing staffing to multinational companies in 26 countries. It offers businesses temporary services, outsourcing, on-site and full-time employment, and consulting services.

The agency defines specific staffing needs of a company and coordinates a workforce that meets its daily business requirements. The contracted firm can then stay focused on its core business and not worry about employee relations. This cost-effective model is popular among many companies.

Kelly Services' commitment to diversity has won its recognition as one of the best companies in the field. Its philosophy is one of forming partnerships with companies that are controlled by minority, women, and disabled veterans. The firm's program to assist and advise these business operations has proven successful.

Source: *http://www.Kellyservices.com*

ACTIVITY A LANGUAGE ARTS RECALL

Hyphen Usage

Rule 12.1: Use a **hyphen** to join a compound adjective (one-thought modifier) when it precedes the word it modifies; otherwise, it is not hyphenated.

Example: Job opportunities exist in ***computer-related*** fields. (precedes noun it modifies)

BUT

Job opportunities exist in fields that are ***computer related.*** (does not precede noun it modifies).

The personnel department keeps ***up-to-date*** records on all employees.

BUT

The personnel department keeps employee records that are ***up to date.***

The supervisor and employee will have a ***face-to-face*** meeting to discuss ***on-the-job*** training.

Note: If an adverb that ends in "ly" precedes the adjective, do not hyphenate.

The human resources department is a ***newly formed*** division of the corporation. (no hyphen because "newly" ends in "ly" and is an adverb)

Rule 12.2: Use a hyphen after words in a series of hyphenated adjectives that modify the same noun(s). When there are two hyphenated adjectives, space once after the hyphen not linked with the noun. When there are three or more, use commas to separate them.

Examples: The ***five-*** and ***ten-year*** trends continue to show an increase in service jobs.

We will review our ***one-, three-,*** and ***five-year*** goals.

Rule 12.3: Use a hyphen after *self* when used as a prefix (*self* words are hyphenated).

Example: ***Self-esteem*** improves when an individual gains ***self-confidence.***

Rule 12.4: Use a hyphen to join the word *well* to another word to form a modifier before a noun; otherwise, no hyphen is used.

Example: A ***well-written*** resume will get positive results.

A resume that is ***well written*** will get positive results.

Rule 12.5: Use a hyphen in simple fractions.

Example: Of the 100 questionnaires mailed, only ***one-third*** replied.

Rule 12.6: Use a hyphen to link compound titles containing an "ex" and "elect."

Example: The acting ***ex-secretary*** joined the ***president-elect*** for lunch.

Note: Titles containing "vice" are not hyphenated.

Example: vice president

Rule 12.7: Use a hyphen to link compound titles in which a person or thing has two functions of a dual nature.

Example: The acting ***secretary-treasurer*** attended the ***dinner-dance*** with the ***manager-supervisor.***

Rule 12.8: Use a hyphen to join a prefix to a number or to a capitalized word.

Examples: The fall semester begins in ***mid-September.*** Marilyn was promoted to general manager in the ***mid-1990s*** when she was in her ***mid-30s.***

Number Usage

The following are the most common rules for expressing numbers in business correspondence.

Rule 12.9: *Word Style—Spell out*

- numbers from one through ten
  Fax five copies of the report.
- numbers beginning a sentence
  Fifty applicants mailed their resumes.
- millions and billions as part of a dollar amount

  $10 million $1.5 billion

Note: Never write the word "dollar" after the amount when the dollar symbol ($) is used.

- fractions that stand alone
  one-third

- street addresses beginning with "one" in inside address
  One Park Avenue
- time of day in formal writing
  eight-thirty o'clock

Rule 12.10: *Figure Style—Write as numbers*

- exact numbers over ten
  Fax 20 copies of the report.
- all decimals and percents (unless they begin a sentence)
  The salesperson received 10 percent commission on all sales.

  Note: Write out word *percent* in correspondence.
- time of day when written with a.m. or p.m.
  8 p.m.
- measurements, dimensions, page numbers, house numbers
  68 degrees 3″ × 6″ card
  page 25 101 Broadway
- numbers in compound adjectives
  70-story building, 10-page manual
  35-hour workweek 3-year budget
- whole dollar amounts
  $50 $5,235

  Note: Do not add a decimal point and zeros.
- series of items

  Note: If any number in a sentence is above 10, write all numbers as figures.

  We need to reorder 5 reams of computer paper, 10 disks, and 25 file folders.

Check for Understanding

Directions: Rewrite the sentences below by changing the *underlined* phrases into one-thought modifiers.

Example: The speaker who is *well informed* addressed the conference.
Rewrite: The *well-informed* speaker addressed the conference.

1. The manager has a commute of 40 *minutes* to work each day.

   Rewrite ______________________________

   ____

2. The firm's plan was to hire workers who were *skilled on the computer.*

   Rewrite ______________________________

   ____

3. The union agreed to a retirement plan that was *sponsored by the employer.*

   Rewrite ______________________________

   ____

4. The communications company hired a consultant who was *well known.*

   Rewrite ______________________________

   ____

5. The employee was given a leave of absence for a *short term.*

   Rewrite ______________________________

   ____

Conclusion: Complete the following sentence.

A _______ is used to join a _______ modifier when it _______ the noun.

(Answers are given below.)

Answers to "Check for Understanding"

The manager has a *40-minute* commute to work each day.
The firm's plan was to hire *computer-skilled* workers.
The union agreed to an *employer-sponsored* retirement plan.
The communications company hired a *well-known* consultant.
The employee was given a *short-term* leave of absence.
Conclusion: hyphen, compound adjective, precedes

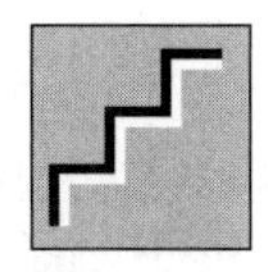

Building English Power: Hyphens and Number Usage

Directions: Punctuate each sentence correctly, inserting or deleting hyphens where necessary. Review number usage and make appropriate corrections, then key each sentence correctly.

1. Employers are interested in self motivated individuals who are well informed about the firms products. (12.3, 12.4, 9.1)
2. The regular conference fee is $20.00 dollars members have a 10% discount. (12.10, 8.1)
3. Applicants should send a follow up letter to thank the interviewer. (12.1)
4. The well known speaker Dr. Robert Reed will present a two day workshop on self assertiveness. (12.4, 7.5, 12.1, 12.3)
5. The applicants interview was scheduled for two p.m. at the company's headquarters at 1 Park Row. (9.1, 12.10, 12.9)
6. The firm budgeted $1.2 million dollars for renovation of its 50 story building. (12.9, 12.10)
7. At the conference the vice president of the firm remarked that the salesforce will be reduced by one third by mid April. (6.5, 12.6, 12.9, 12.8)
8. The new contract includes a three week vacation for employees who have been in the company for 5 years. (12.1, 12.9)
9. Home based work is a good option for part time employment. (12.1)
10. The 12 page manual will give you detailed step by step instructions for the new computer software. (12.1, 7.7, 12.2)
11. Seventeen and eighteen year olds are encouraged to find volunteer internships to gain hands on experience. (12.10, 12.2, 12.1)
12. Entry level jobs have become more demanding with the need for higher technical skills. (12.1)
13. The president of the company is well-known in the electric utility industry. (12.4)
14. Some individuals prefer the freedom of a freelance job rather than a traditional 8 hour position. (12.1, 12.10)
15. 20 applicants were interviewed for the 3 positions as paralegals. (12.9)

ACTIVITY B BUILDING WORD POWER

The words below are commonly used in the field of human resources. From this list, select the word or phrase that correctly completes the statements. Refer to the Personal Dictionary, if necessary.

assertiveness	internship
assess	modulated
commitment	perusal
crucial	preliminary
human resources	prospective
	teleconferencing

________ 1. The administrative assistant has a well-________ voice when he speaks on the telephone.

________ 2. The meeting between the San Francisco and Tokyo offices was conducted by ________ so that both parties could discuss the issue via television.

________ 3. The job applicant was interviewed by a staff member of the ________ department.

________ 4. An employee who is interested in promotions within a firm should demonstrate ________ to the company.

________ 5. In the highly competitive field of sales, self-________ is an important trait for a successful career.

________ 6. The manager will ________ the present procedure used in interviewing applicants.

________ 7. In today's job market, some people accept a(n) ________ to work with no salary for possible future full-time employment.

________ 8. A(n) ________ meeting was held by staff to prepare for the president's press announcement.

________ 9. A person's integrity and honesty are ________ traits for conducting business transactions.

________ 10. The salesperson called the ________ client regarding the future order.

________ 11. The application is enclosed for your ________.

On the Alert

Attention Line

Postal regulations require that the attention line appears as the first line of the address on the envelope; use the same format for the inside address of the letter. Use "Ladies and Gentlemen:", "Gentlemen:", or "Ladies:" for the salutation.

Example: ATTENTION HUMAN RESOURCES DEPARTMENT
ANCHOR CORPORATION
55 PINE STREET
EDISON NJ 08032-3387

Ladies and Gentlemen:

ACTIVITY C INDUSTRY PERSPECTIVES: RECRUITING IN TODAY'S JOB MARKET

Directions: Using appropriate revision marks, proofread and key the following report titled RECRUITING IN TODAY'S JOB MARKET.

Rule Reference

The changing workplace and the popularity of the Internet has had major affects on job search and recruitment. Other than the traditional way of looking
2.1 for a job, alternative strategies exists for both the employer seeking qualified workers and the individual searching for a job.

What are some things you need to know about the job
6.3 search process. After deciding on a career path you will need to do some preliminary research about the firms in this field. Having knowledge of the company where you are seeking employment can be crucial in getting a job offer. An excellent source of information is the web site *http://www.hoovers.com*.

6.3 Before writing the resume you should prepare a list of your skills, educational background, work experiences
7.1 and other related activities. Then you will decide how you plan to pursue the job search. The traditional way of searching for a job in the newspaper is being overshadowed by other alternative means. One of the methods used most is the Internet.

INTERNET JOB RECRUITING

Employers are using various channels to list job openings. Special web sites similar to a newspaper
2.1 listing is available. Some of the popular sites are *http://www.hotjobs.com*; search engines, like yahoo, also post job openings. Many companies that post openings have there own web sites. The job seekers then submit their resumes online or fill out application forms on the Internet.

Another electronic service is the Internet provider that acts, for a fee, as an online employment agency. It maintains current databases of recruiters who have job openings and resumes of job seekers. Recruiters are supplied with the resumes that match their needs. Job seekers are matched only with those openings in which they qualify. This method of filling positions is faster and more efficient.

PERSONAL INTERVIEW

12.1 In the present type of job market, a face to face interview only takes place after an employer has
6.3 determined which candidates qualify for the opening. At the interview the applicant needs to be well-prepared,
7.1 be knowledgeable about the firm make a good appearance, and have a positive attitude.

JOB FAIRS

Job fairs are popular throughout the country. It gives a group of employers in a geographic area or in a specific industry the opportunity to meet with perspective applicants. At the same time, the job seekers can evaluate the existing job market, obtain job
7.1 information on a number of firms and interact with company recruiters.

SUMMARY

12.1 The job search is a time consuming and challenging process. The best advise is to be prepared and to be
12.3 self confident that you are the one for the job.

Did you find these errors?

5 hyphens 6 commas 4 word usage 2 subject–verb agreement
1 punctuation

Cover Letter

The **cover letter** or job application letter introduces the prospective employer to the applicant's resume. It needs to attract the reader's attention. The letter does not repeat the information on the resume. It should highlight the applicant's qualifications for the job as well as the person's strong points.

A cover letter can be written to a specific firm or organization that has posted an opening (solicited) or to firms that you believe would be a good fit for your skills and career plans (unsolicited). You should always attempt to address the letter to a specific individual.

In the opening of a solicited cover letter, refer to the source of information whether it is a newspaper ad, Internet posting, or contact. In an unsolicited letter, specify the type of work you are interested in doing.

The body of the letter should highlight your qualifications that meet the needs of the job and benefit the organization. You may also refer to academic or technical courses that you have taken and explain how they can help your work. Also, indicate how your previous work experience will be valuable to your prospective employer.

A request for an interview should appear in the closing.

Remember that in essence the cover letter is a sales letter. It should be effective and professional looking.

TIP: Internet Sources of Job-Seeking Information

The following web sites are excellent sources of information on career opportunities, job listings, and resume writing:

http://monster.com

http://www.careerpath.com

http://www.hotjobs.com

http://www.careerjournal.com

http://www.careerbuilder.com

Model Cover Letter

Designed letterhead

LAURIE WHEELER
500 River Lane
Riverdale, NY 10471

(Current Date)

Addresses specific person

Mr. Joseph Engers
Human Resources Manager
EGR Realty Corporation
One Park Avenue
New York, NY 10057-3200

Dear Mr. Engers:

Identifies source and position

Your advertisement in the May 1 edition of the *NEW YORK TIMES* for an administrative assistant interests me.

Relates education to job requirements

I recently graduated with an A.S.S. degree from Bronx Community College with a major in Office Information Systems. This curriculum gave me the necessary communications, computer, and interpersonal relationship skills that will prove valuable in the position that is available in your firm.

Refers to firm's reputation

Emphasizes qualifications

EGR Realty Corporation has a fine reputation for its innovative approach in the real estate business. While a student, I worked at Riverdale Realtors where I became very attracted to a career in the dynamic, challenging field of real estate. You will find that I am a highly motivated person who can deal effectively with clients and staff.

Refers reader to resume

Shows interest in job

My resume is enclosed for your perusal. I would appreciate meeting with you to discuss my qualifications and possible employment with your company. You may reach me at (718) 220-7832.

Sincerely yours,

Laurie Wheeler

ACTIVITY D EDITING POWER

Task A: Edit, format, and proofread the following unarranged letter, then key it correctly. Remember to insert paragraphs.

You, as the applicant, should send this cover letter to the attention of the Human Resource Department, Anchor Corporation, 55 Pine Street, Edison, New Jersey 08032-3387.

Errors
- commas
- hyphen
- capitalization
- underscore
- typographical
- spelling
- verb tense

Please consider me as an applicant for the position of sales representative which you advertised in The New York Times on sunday May 16. I have chosen your company because it is a well respected organization with a reputation for supporting employees in their professional developement. As shown in my enclosed resume you will find that my educational background has provided me with an excellent foundation for the position for which I am applying. I am skilled in the use of Microsoft office software, including Word Excel and PowerPoint. My position as an administrative assistant has enable me to develop my interpersonal and communication skills which are essential in a sales position. I will bring to the job a willingness to work and an eagerness to learn. May I have the opportunity to discuss employment possibilitities with you? I would like to schedule an interview at your convenience. My telephone number is (student's telephone number).

Writing Pointers

Interview Follow-up Letter (Refer to Activity D, Task B for example of letter content)

- Send letter immediately after interview.
- Express appreciation for the interview.
- Remind the interviewer of your interest in the job.
- Reemphasize a positive point about yourself.
- End with a forward-looking statement.

Task B: Edit, format, and proofread the following handwritten letter for all errors; key it correctly. It is a follow-up thank-you letter for the job interview with Ms. Michelle Dowling at Anchor Corporation, Human Resources Department, 55 Pine Street, Edison, NJ 08032–3387.

Thank you for interviewing me on Thursday May 10 for the sales representative position with your firm. The position appears interesting and challenging. I have always admired Anchor Corporation for it's forward looking approach in the field of sales. I am confident that I possess the talent skills and commitment that you require of your employees. You will find me an energetic member of any team effort. I am eager to join your staff and I hope to hear from you soon.

ACTIVITY E WRITING PRACTICUM

CHOOSE OPTIONS FOR REVISION OF A COVER LETTER

Situation: Mrs. Dorothy Frank told her neighbor's daughter, Michelle Yager, a second-year college student, that there are five summer internships available in the City Manager's office where she works. Michelle composed the following letter to the human resources director. You have been asked to read the letter and improve the writing.

Directions: Read the entire letter, which consists of a salutation, closing, and numbered sentences. Focus on one sentence at a time and select the best option from the choices given. After completing this task, key the letter.

(1) Dear Mr. Keith Shane:

(2) I am writing to you at the suggestion of Mrs. Dorothy Frank. (3) Dorothy said there are 5 summer-internships available in your office and I am most interested in applying for 1 of them. (4) At Queens college I am majoring in municipal government.

(5) Could you please send an internship application form to the above address and include any requirements you might have for obtaining one of them. (6) Enclosed is a self addressed stamped envelope for your convenience.

(7) If necessary I can provide 3 letters of reference. (8) I am available any day after 2 p.m. for an interview. I look forward to hearing from you. (9) Yours Sincerely,

Revisions

(1) A. Dear Mr. Shane:
B. Dear Sir:
C. Dear Keith,
D. Original salutation
(1) Choice _______

(2) A. I am writing to you at the suggestion of Dorothy Frank.
B. I am writing you at the suggestion of my neighbor.
C. I am writing to you at the suggestion of my neighbor, Mrs. Dorothy Frank, one of your employees.
D. Original sentence
(2) Choice _______

(3) A. Dorothy said there are 5 summer internships available and I am most interested in applying for one of them.
B. She informed me that there are five summer internships available in your office, and I am most interested in applying for one of them.
C. She said there are five summer internships available in your office and I am most interested in applying for one of them.
D. Original sentence
(3) Choice _______

(4) A. I am at Queens College and majoring in municipal government, a field most fascinating.
B. At Queens College, I am majoring in municipal government, a career field that I find fascinating.
C. At Queens College, I am majoring in municipal government. This is fascinating to me.
D. Original sentence
(4) Choice _______

(5) A. Could you please send an internship-application form to the above address and include any requirements necessary for obtaining one of them?
B. Please send an internship application form to me at the above address and include any other requirements needed in applying for the internship.
C. Please send me an Internship-Application form to the above address and include any suggestions you might have in obtaining one of them.
D. Original sentence
(5) Choice _______

(6) A. Enclosed is a self addressed stamped-envelope for your convenience.
B. Enclosed is one self addressed stamped envelope for your convenience.
C. I am enclosing a self-addressed, stamped envelope for your convenience.
D. Original sentence
(6) Choice _______

(7) A. If required I can send three letters-of-reference.
B. If necessary, I can provide three letters of reference.
C. I can get 3 letters of reference if required.
D. Original sentence
(7) Choice _______

(8) A. I am available any day after 2:00 pm for an interview, and am looking forward to hearing from you.
B. I am available any day after two o'clock for an interview. Thank you for your assistance.
C. I am available any day after 2:00 pm for an interview and thanks for considering my application.
D. Original sentence
(8) Choice _______

(9) A. Sincerely yours,
Michelle Yager
B. Appreciatively yours,
Michelle Yager
C. Truly yours,
Michelle Yager
D. Original closing
(9) Choice _______

(10) For improved writing, select one of the following options.

A. Rearrange sentences in the following order: 2, 4, 3, 5, 6, 7, 8.
B. Combine paragraphs 1 and 2.
C. Key sentences 5, 7, and 6, in that order, into a second paragraph. Other sentences remain in the same order.
D. Keep letter in original order.

(10) Choice ________________

Key the letter with the choices and revisions you have made.

Resumes

A **resume** is a marketing tool that outlines a person's skills and experience. It presents an image you want employers to see. The resume should focus on accomplishments, not past job duties. The main goal of a resume is to obtain a job interview. A resume should include the following details.

Contact Information: At the top of the resume, the applicant's full name, address, phone number, and e-mail address should appear.

Objective: The recruiter should get an immediate sense of who the applicant is. Stress what you can do for the company. You can make a general statement if the resume is not targeted for a specific firm or position.

Experience: List job experiences chronologically; your most recent job comes first. Emphasize important responsibilities. Be specific; use action verbs.

Skills: Focus on skills employers want.

Education and Honors (if applicable): The latest academic enrollment information should be written first. This section can appear as the first major category, especially if the individual is a first-time applicant.

Other Activities: Include any activities that may be related to your career or that show involvement in the community or other affiliations.

Categories of Resumes

There are three categories of resumes: **chronological, functional,** and **combination.** The applicant should use the one that is best suited for the job search. The *chronological* resume outlines the job history of the applicant and starts with the most recent position. This category is popular with candidates who have extensive work experience.

The *functional* resume stresses the skills and educational accomplishments of the applicant and is good for beginning job seekers.

The *combination* resume takes the best from both types, including work experience as well as skills, competencies, and educational information.

Writing Pointers—Resumes

Guidelines to follow when preparing resumes:

- List strongest skills or abilities that make you a good candidate.
- Describe accomplishments emphasizing the results that benefited the past employer.
- Focus on qualifications that reflect the needs of the hiring company.
- Eliminate wordy statements; be specific.
- Avoid spelling errors and poor grammar.
- Omit unrelated personal information.
- Include contact information.
- Include soft skills (team player, flexibility, self-motivated).
- Use an easy, readable format.

Electronic Resumes

Resumes are often scanned electronically by companies. Many job search engines, specialty job sites, job hunting companies, and newspapers have a database tool in which the software reads the resume and stores it in a database. Prospective employers can then peruse the resumes. If the software cannot read the resume, it is rejected. Therefore, a *scannable resume must be modified from the traditional, printed version.* Make sure you do the following:

- Provide a list of keywords, preferably nouns, that are descriptive of your skills, abilities, education, and job titles. These words are matched with the employer's list. This section follows the heading in the resume.

- Avoid italics, bold face, underlines, bullets.
- Do not use abbreviations and acronyms.
- Use a font that produces a clear, plain text copy.
- Leave white space to separate sections.
- Left-justify the document; eliminate columns.
- Print on high-quality white or light-colored paper.

An **online resume** can be sent as an attached file or pasted text in body of an e-mail. For security reasons, you may want to remove the home address and telephone number from the heading; keep only the e-mail address.

Model Resumes

The two model resumes that appear on pp 334–336 represent a beginning job applicant. The first example, a printed version, is a combination resume. The emphasis, however, is on the educational background and the applicant's skills.

The second version is an example of how the traditional, printed version is adapted to a scannable or e-mail resume.

ACTIVITY F WRITING POWER: COMPOSING CORRESPONDENCE

CASE STUDY 1. LETTER OF APPLICATION TO BUSINESS

After reading the case, respond to the questions in "Organize Your Ideas," then compose the letter.

Case: One of your instructors recommended that you send a letter with your resume to the well-known accounting firm Forrest & Smith, 55 Midland Drive, Chicago, IL 60607–8935 for possible employment in an entry-level position. You are graduating with a two-year degree in accounting from your local community college. You have completed a three-month internship with a local accountant, where you assisted with financial records and payroll. Don't forget to indicate your computer skills.

(Note: Case Study 1 continued on page 336)

Model Resume Printed Version

Laurie Wheeler
500 River Lane
Riverdale, NY 10471
(718) 220-7832
lauriew@yahoo.com

Uses general objective

CAREER OBJECTIVE

To obtain a challenging position as an administrative assistant with a multinational firm that is a leader in its field.

EDUCATION

Emphasizes education and skills; highlights related courses

Bronx Community College of CUNY
Bronx, NY 10453-3101
Associate of Applied Science degree, Office Information Systems Major, GPA 3.5 (2005)

Related courses include:

Accounting	Multimedia Theory
Business Communications	Principles of Interpersonal Communications
Microcomputer Applications	Web Page Design

Skills:

Microsoft XP Professional, Microsoft Office (Word, Excel, PowerPoint, and Access)

Honors:

Dean's List 2003–2005

J.P. Murphy's Award for Volunteerism

Bold side headings and bullets for clarity

WORK EXPERIENCE

Bronx Community College, Bronx, NY
Student Assistant to Information Systems Department
(school session 2003–2005)

Uses action verbs

- Worked closely with faculty with filing, faxing, and copying documents
- Inputted student information into a database
- Assisted faculty in processing course registration
- Interacted with students on departmental matters

Riverdale Realtors, Riverdale, NY (summers and part-time 2003–2004)

- Inputted potential clients' information into a database
- Recorded and forwarded telephone messages to real estate agents
- Greeted customers
- Assumed responsibility for running office when Realtors were off-site

Lists activities that show involvement

ACTIVITIES

- Volunteered to tutor lower-grade students in reading skills
- Volunteered to help children with special needs in recreational activities

Model Electronic or Scannable Resume

Online: addresses and telephone numbers may be omitted

Laurie Wheeler
500 River Lane
Riverdale, NY 10471
(718) 220-7832
lauriew@yahoo.com

Lists applicable keywords

KEYWORDS
Administrative Assistant. Microsoft XP Professional. Microsoft Office. Word Excel. PowerPoint. Access Communication Skills. Organizational Skills. Interpersonal Relations. Volunteer. Student Assistant. A.S.S. degree, Bronx Community College. Office Information Systems Major.

Avoids bold headings

CAREER OBJECTIVE
To obtain a challenging position as an administrative assistant with a multinational firm that is a leader in its field.

Write out degree title to avoid confusion

EDUCATION
Bronx Community College of CUNY, Bronx, NY
Associate of Applied Science degree, 2005
Office Information Systems Major. GPA 3.5

Related courses include:
Accounting, Business Communications, Microcomputer Applications, Multimedia Theory, Principles of Interpersonal Communications, Web Page Design

Uses white space for clarity

Skills:
Microsoft XP Professional, Microsoft Office, Word, Excel, PowerPoint, Access

Honors:
Dean's List 2003–2005
J.P. Murphy's Award for Volunteerism

WORK EXPERIENCE
Bronx Community College, Bronx, NY
Student Assistant to Information Systems Department, school session 2003–2005

- Worked closely with faculty with filing, faxing, and copying documents
- Inputted student information into a database
- Assisted faculty in processing course registration
- Interacted with students on departmental matters

(continues)

Model Electronic or Scannable Resume—Concluded

Riverdale Realtors, Riverdale, NY, summers and part-time 2003–2004

- Inputted potential clients' information into a database
- Recorded and forwarded telephone messages to real estate agents
- Greeted customers
- Assumed responsibility for running office when Realtors were off-site

ACTIVITIES
Volunteered to tutor lower-grade students in reading skills
Volunteered to help children with special needs in recreational activities

CASE STUDY 1. CONTINUED

Organize Your Ideas

1. What would be a strong opening statement to attract the reader's attention?

____

____

2. What important points would you emphasize in the body of the letter regarding your individual strengths?

____

____

3. What would you say in the closing statement to get an immediate response?

____

____

CASE STUDY 2. LETTER OF APPLICATION IN ANSWER TO NEWSPAPER ADVERTISEMENT

After reading the local newspaper advertisement that follows, which appeared in *The Daily Press,* answer the questions in "Organize Your Ideas" and draft a letter of application to Box MPM, 100 East 43 Street, New York, NY 10017. Use "Ladies and Gentlemen:" as the salutation.

Case:

ADMINISTRATIVE ASST
ENTRY LEVEL OK

National weekly newspaper seeks a highly organized individual for its Customer Relations Department. Applicant must be good with people. Verbal and typing skills a must. Computer knowledge essential. Good entry-level position for motivated individual.

Organize Your Ideas

1. What would you write in your opening statement to stimulate the reader's interest in you as a potential candidate for the job?

____

____

2. What specific details would you highlight in the body of the letter to focus on your strengths in reference to the position?

____

____

3. What would you say in your closing statement to get a favorable response?

____

____

CASE STUDIES 3 AND 4. COVER LETTER AND RESUME OF STUDENT

The writing activities in this chapter have focused on the types of correspondence that you will compose during your job search. In this exercise, you will draft a cover letter and resume for a position that you personally would seek. To help you write an effective resume, the model previously shown can serve as a guide when you prepare your own. This sample resume contains the parts that are found in all resumes; however, the format might differ.

Case: Select a job for which you might wish to apply that was advertised in your local newspaper. Write a cover letter and resume based on your qualifications and background.

CASE STUDY 5. INTERVIEW FOLLOW-UP LETTER

A thank-you note after an interview helps you stand out against other applicants. If you are interested in a position, the follow-up letter will remind the interviewer of your meeting. This is an important final step in the interview process.

Case: Write a follow-up letter after an interview for a position as an administrative assistant that you had with Sandra Alexander, Human Resources representative for Intercontinental Electronics Corporation. This is a highly respected telecommunications company that has a reputation as a global leader in the field. They are located at 10037 Country Road, San Jose, CA 95100–5321.

ASSESSMENT: CHAPTER 12

English Power

Directions: *In the following sentences, insert punctuation and correct number usage where needed and proofread. Then key each sentence.*

1. If you want a successful corporate career you are expected to be self confident, well informed and able to work on a team.
2. The 35 page annual report features the companys overseas operations.
3. Only one third of the individuals contacted responded to the survey on self employment.
4. Upper level managers require years of experience and advanced education.
5. The new contract includes a four week vacation for employees who have been with the firm for 5 years.
6. We offer a well designed course on decision making techniques for managers.
7. This one day seminar is guaranteed to provide you with up to date supervisory methods that you can use on the job.
8. John Smith a 40 year old engineer with 2 children has been out of work for 3 months.
9. We budgeted $3 million dollars for renovation of our 20 story building.
10. After a six month course the applicant took a 4 hour exam in computer skills.

Editing Power

Directions: *Edit, format, and proofread the following unarranged form letter for errors; key it correctly.*

```
Dear Participant: Are you apprehensive about reentering
the workforce? Do you feel that you are not progressing
in your present job? On Saturday June 1 the Office of
career Development will hold an important all day
seminar for individuals like you. Work shop sessions
will be led by well respected knowledgeable
representatives from a variety of areas. Among the
topics to be covered are the following self assessment
communication skills and interviewing techniques. These
hands on meetings will be limited to fifteen individuals
so that you will be able to interact with the workshop
representatives. Time will be devoted to answer your
personal questions. The fee is only $75.00 if you
preregister by mail otherwise the fee at the door will
be $90.00. We hope that you will join us for a self
fulfilling day. Please call or fax us your registration
at (914) 553-4203. Sincerely, Beth Morales, Director
```

Writing Power

Directions: *After reading the newspaper ad below, draft a letter of application to the attention of the Human Resources Department, Basic Computer Corporation, 10 West Main Street, Salt Lake City, UT 84109–3322.*

Marketing Support Representative. Basic Computer Corporation, one of the largest global computer suppliers, is expanding its operations. We are offering an opportunity for a recent graduate interested in a career in sales and marketing in the computer industry. We are seeking responsible candidates who will be involved in customer contact, order placement, and problem solving. Computer knowledge is a must

Team player: One who unites other toward a shared destiny through sharing information and ideas, empowering others, and developing trust.

Dennis Kinlau

Appendix A

Answer Keys by Chapter

CHAPTER 1: BANKING ON GOOD SENTENCE STRUCTURE

Activity A Building English Power: Sentence Structure

Exercise 1

1. I (doesn't make grammatical sense)
2. C, We, invite
3. C, charges, do appear
4. I (no verb)
5. I (no verb)
6. C, you, need
7. I (no verb)
8. C, you, have
9. C, bank, requested
10. I (doesn't make grammatical sense)
11. C, advisor, was
12. C, deposit, is encouraged
13. C, bank, will inform
14. I, (doesn't make grammatical sense)
15. C, banks, require

Exercise 2 *(Answers may vary.)*

1. Your monthly itemized billing statement lists all charges.
4. Free additional credit cards are available for your family.
5. No annual fee is required for the first year.
7. Private loan companies that offer personal loans and mortgages compete with banks.
10. If you pay your credit card balance in full each month, you will avoid interest charges.
14. A penalty fee is charged to the depositor's account for tracing a lost check.

Building English Power: Capitalization

1. You must reply by September 1 to receive a free desk calendar.
2. Mastercard and Visa are the most popular bank credit cards.
3. New York Bank will be closed on Saturdays during July and August.
4. Most savings accounts are insured by the FDIC, a government agency known as the Federal Deposit Insurance Corporation.
5. What did you think of Mr. Baker's speech in San Francisco on banking reforms?
6. The Federal Home Loan Mortgage Corporation announced that mortgage rates are on the rise.
7. Eastern Trust offers an IRA that has no annual fee.
8. A CD account offers the investor a higher rate of interest.
9. The new computer software, known as Bank Express, will allow users to keep personal and financial records up to date.
10. On Thursday, December 1, the Royal Bank will introduce its new service for investors.
11. Dr. Alan Smith will speak on global investments at the Orlando Convention Center in Florida on Saturday, July 8.
12. ATMs are located in many shopping malls in the United States.
13. The American Bank will sponsor the next World Cup USA.

14. *Ⓟathways to Ⓕinancial Ⓢecurity* has made *Ⓣhe Ⓝew Ⓨork Ⓣimes* best-seller booklist. ("The" is part of the title.)

15. Ⓣhe Ⓔuropean Ⓑank will expand its offices to Ⓢingapore and Ⓗong Ⓚong.

Activity B Building Word Power

1. insufficient funds
2. electronic banking
3. global
4. mortgage
5. authorization
6. transactions
7. specialists
8. accrued
9. co-applicant
10. network
11. IRAs
12. ATM
13. collateral
14. FDIC
15. CDs
16. debit card

Activity E Writing Practicum

Exercise 1 Writing in Complete Thoughts *(Answers may vary.)*

1. When a small Midwestern bank failed, the FDIC assured depositors that their money was insured.
2. A retirement savings plan is one of the features offered by a full-service bank.
3. Mr. Sanders received a free gift from his bank when he opened a new savings account.
4. One of the restrictions in a CD account is that there is a penalty for early withdrawal of funds.
5. As interest rates dropped, many customers of the bank chose to invest their money in mutual funds.
6. Loaning money to individuals and firms is one of the ways in which banks make money.
7. The slow teller at the second window was much less efficient than the teller at the last window.

8. If mortgage rates continue to increase, prospective homeowners will not be able to afford purchasing a home.
9. The letter from Mr. Robinson, the bank president, explained the bank's new policy on personal loans.
10. Keeping accurate records of expenditures is one of the advantages of a checking account.

Exercise 2 Combining Sentences

1. ATMs are a convenient way for busy persons to conduct banking transactions 24 hours a day.
2. The bank has a special loan rate of 7 percent on American-made cars.
3. A certificate of deposit is a term savings account that pays a higher rate of interest and carries an early withdrawal penalty.
4. Many depositors have online accounts on the Internet to download statements and to pay their bills.
5. You can open a basic checking account for as little as $25 without a minimum balance.

Exercise 3 Developing Meaningful Paragraphs

1. Banking in America has changed dramatically over the past decade. K
2. Banks have responded to changing needs of the consumer in the local communities. K
3. The electronic age has introduced ATMs, direct deposits, and computer transactions, which improved banking services. K
4. Customers now receive fast and efficient service without long waiting lines. R
5. Also, with electronic banking systems, customers can transact their financial business from any area of the country. K
6. Also, a large number of banks have merged to offer additional services that facilitate handling global transactions. K
7. They are open some evenings, Saturdays, and Sunday mornings. K
8. With these changes, banks are prepared to meet the challenges of the next millennium. R

- Order of Sentences: 1, 3, 4, 5, 2, 7, 6, 8

Rewritten Paragraph: Banking in America has changed dramatically over the past decade. The electronic age has introduced ATMs, direct deposits, and computer transactions, which improved banking services. Customers now receive fast and efficient service without long waiting lines. Also, with electronic banking systems, customers can transact their financial business from any area of the country.

Banks have responded to changing needs of the consumer in the local communities. They are open some evenings, Saturdays, and Sunday mornings. Also, a large number of banks have merged to offer additional services that facilitate handling global transactions.

With these changes, banks are prepared to meet the challenges of the next millennium.

Exercise 4 Writing Opening and Closing Sentences in Letters

Opening Sentences

1. B (Question attracts reader's attention.)
2. A (Clear purpose stated)
3. B (Informative)
4. B (Goodwill approach)

Closing Sentences

5. B (Suggests action to be taken)
6. B (Positive forward-looking statement)
7. A (Offers a friendly invitation)

CHAPTER 2: ENSURING SUBJECT-VERB AGREEMENT

Activity A Building English Power: Subject-Verb Agreement

1. plan, protects
2. procedures, help
3. club, has
4. you, ask
5. union and management, have
6. industry, faces
7. policy, was
8. contract, includes
9. Michael and friend, are
10. course, qualifies
11. you, become
12. damage and theft, are
13. records, indicate
14. team, has
15. insurance and coverage, are

Activity B Building Word Power

1. catastrophic insurance, umbrella insurance
2. phasing out
3. escalating
4. cafeteria-type medical insurance, option
5. bodily injury insurance
6. property damage insurance
7. deductible
8. premium
9. subsidized
10. accelerated
11. alternative
12. liability insurance

Activity E Writing Practicum

Exercise 1 Writing Opening and Closing Sentences

Note: Choices should be based on clarity and conciseness.

Opening Sentences

1. A (identifies writer and states purpose of letter)
2. C (clear statement, not wordy)
3. C (to the point)
4. A (concise and clear)
5. B (expresses pleasure in fulfilling request)

Closing Sentences

6. A (brief statement of appreciation)
7. A (concise statement; builds goodwill)
8. B (concise statement encouraging future business)

Exercise 2 Writing for Correct Sentence Structure

You could save hundreds of dollars by insuring your automobile with us. Our operating expenses are among the lowest in the industry, and we pass our savings on to our customers.

We have additional money-saving discounts for safe drivers, auto security systems, and multi-car coverage. Our quality service includes 24-hour access to our auto insurance specialists.

Use our toll-free number to get a rate quote; there is no obligation.

CHAPTER 3: INVESTING IN SUBJECT-VERB AGREEMENT

Activity A Building English Power: Subject-Verb Agreement

1. Foreign demand for American technology is increasing.
   Plural Form: demands, are increasing
2. The number of stockholders is demanding more communication from corporate executives.
   Plural Form: A number, are
3. Neither the employee nor the manager agrees with the arbitrator's decision.
   Plural Form: employees, managers, agree
4. There is a concern about the company's financial condition.
   Plural Form: are, concerns
5. The report in which we compare multinational corporations appears in several business magazines.
   Plural Form: reports, appear
6. The enclosed pamphlet from the brokerage firm gives information on a variety of mutual funds.
   Plural Form: pamphlets, give
7. Each of the brokers predicts an upswing in the stock market.
   Plural Form: All, predict
8. A seminar on financial planning is offered by the local bank.
   Plural Form: Seminars, are offered
9. There is a tax-free municipal bond that pays a high dividend.
   Plural Form: are, bonds, pay

10. Everyone agrees that individuals should have a savings plan.
    Plural Form: All, agree
11. Either a U.S. savings bond or a treasury note is a secure investment.
    Plural Form: bonds, notes, are
12. Each financial decision impacts your net worth.
    Plural Form: (ALL) decisions, impact
13. The stock price of the electronic company has increased during the past year.
    Plural Form: prices, have increased
14. The fund, which invests in diversified stocks, has a high rate of performance.
    Plural Form: funds, invest, have
15. Neither the bank nor the investment firm was willing to underwrite the project.
    Plural Form: banks, firms, were willing

Activity B Building Word Power

1. bonds
2. securities
3. stocks
4. prospectus
5. investments
6. investment portfolio
7. mutual fund
8. stockbroker
9. glossary
10. multinational
11. proxy
12. web site
13. blue chip stock(s)

Activity E Writing Practicum

Exercise 1 Rearranging Sentences for Continuity of Ideas

1. a, c, b, d

The use of computers to keep abreast of financial information is becoming increasingly popular.

With the Internet, an investor can communicate with investment specialists, retrieve securities information, and refer to financial databases. Online computer networks are making conventional ways of researching information as obsolete as using a typewriter. All you need is a personal computer, a modem, and Internet access to keep the information superhighway at your fingertips.

2. b, a, c, d

In response to your advertisement placed in the summer job mart at River Community College, I would like to apply for the position.

Last summer, I had experience working as an intern for a small investment firm in my hometown. I was fascinated by the challenges of the brokerage field.

Please send me an application for your management trainee program.

3. d, c, a, b

You will be pleased to know that we are introducing an automatic monthly investment plan.

With this plan, you authorize us to transfer a fixed amount each month from your bank account to your mutual fund account. There are no fees for this service, and you can terminate the plan at any time.

If you have any further questions, please call your account representative.

4. d, c, b, a

Thank you for contacting Investment Services Corporation.

Our firm offers a wide range of investment plans tailored to the needs of our clients. All of our investment consultants are professionals who are knowledgeable in the securities market.

We have enclosed a brochure that explains our services; feel free to contact one of our experienced representatives for further help.

5. b, a, c, d

We have enclosed your free copy of *Financial Review*.

As an experienced investor, you will find this monthly newsletter an incredible source of information on all types of investments. Each month a prominent business professional writes about a financial topic of interest to our readers.

Please take time to read this valuable pamphlet and don't hesitate to send in your subscription.

Exercise 2 Inappropriate Phrases

1. Many
2. According to
3. Now
4. Because
5. Should have
6. Because
7. Because
8. Ought not to
9. Must
10. Somewhat

CHAPTER 4: TELECOMMUNICATING WITH VERBS

Activity A Building English Power: Verb Tense

1. office, introduced, simple past
2. machines, have decreased, present perfect
3. interest, has increased, present perfect
4. corporation, will have introduced, future perfect
5. editor, has used, present perfect
6. senior citizens, will have outnumbered, future perfect
7. control, has been discussed, present perfect
8. you, have, simple present
9. task force, has been organized, present perfect
10. manager, will be hired, simple future
11. company, had planned, past perfect
12. executives, carry, simple present
13. directors, had discussed, past perfect
14. we, had demonstrated, past perfect
15. bank, has encouraged, present perfect

Activity B Building Word Power

1. videoconferencing
2. facsimile equipment (fax)
3. modem
4. laptop computers
5. telecommunications
6. electronic mail (e-mail)
7. notebook
8. monitor
9. mouse
10. configuration
11. software
12. Internet
13. protocols
14. networking

Activity E Writing Practicum

Exercise 1

DATE: (Current Date)
TO: Branch Offices
R. J. Tompkins Corporation
FROM: Ray DeJesus
SUBJECT: Telenetwork System

We wanted to alert all of our branch offices that we converted our telephone lines to a new computerized system on September 5.

This allowed for several improvements over our old system. Each new desk unit consisted of a modular console that included fax, voice mail, and three-way conferencing features. Executive offices had hookups for videoconferencing and taping facilities. We were confident that our interoffice communications benefited from this change.

The plan was to install the system over the Labor Day weekend so that the business day had few interruptions. We appreciated your cooperation in this matter.

Exercise 2

The Internet has introduced a new era in communications. Through a computer network, friends, business colleagues, and customers have interacted and have shared information instantly with one another. The Internet has been as popular as using a cellular phone or faxing a message. It has been the focus of the computing world.

Computer networks have changed the way business has handled its transactions. Information has been globalized and resources have been available from all over the world. The following services have been available to the Internet user: online shopping, banking transactions, checking status of orders, researching databases, voice messaging, and exchanging e-mail.

The Internet has grown rapidly. Protocols for this online computer network have not limited the type of users. An increasing number of large corporations, small businesses, and individual subscribers have joined the Internet for their means of communication.

CHAPTER 5: LOTS OF COMMAS IN REAL ESTATE

Activity A Building English Power: Comma Conjunction

1. The home improvement business has become a major industry, and future trends indicate a growing market.
2. The custom-built homes are designed by a well-known architect, but the prices are moderate.
3. The residential community is located near the highway and is convenient to major shopping malls. (**C**orrect)
4. The five acres of land will be sold to individual buyers or to a real estate developer. (**C**orrect)
5. Mortgage rates have been low during the last year, and that will affect future sales.
6. Office space rents by the square foot and is quite costly in desirable business districts. (**C**orrect)
7. Waterfront property has become increasingly expensive, but buyers are always interested in purchasing such real estate.
8. The tenant owes three months' rent, and that is the reason the landlord will take legal action.
9. Congratulations on your election to the governance committee of the building, and the Board thanks you for your contributions to good management.
10. The architect's design for the neighborhood park has been completed, and construction should begin in two months.
11. The land will be sold by the acre and parceled out in smaller lots. (**C**orrect)
12. The property has easy access to the downtown area, and that site is located right off Exit 23 on the interstate highway.
13. The court ruling prohibits the use of wetlands for commercial development, but the firm will appeal the case.
14. A day-care center will be built in the new corporate park, and it will be available to employees and the community.
15. The Association of Realtors will hold its annual conference in June, and we urge individuals who are interested in the field of real estate to attend the meeting.

Activity B Building Word Power

1. references
2. town code
3. Debris
4. periodically
5. index
6. principal
7. lease
8. security deposit
9. corporate park
10. Realtors
11. mortgage
12. condominium
13. cooperative
14. prorating
15. delinquent

Activity E Writing Practicum

Exercise 1 Adjectives and Adverbs

1. scheduled (adjective)
2. completed (adjective)
3. interested (adjective)
4. densely (adverb)
5. satisfactorily (adverb)
6. required, revised (adjective, adjective)
7. leased (adjective)
8. newly (adverb)
9. recently, acquired (adverb, adjective)
10. experienced, fairly (adjective, adverb)

Exercise 2 Comparative and Superlative Forms

1. largest
2. more
3. An
4. most
5. more
6. most
7. a
8. most
9. an
10. an

Exercise 3 Correct Word Choice

1. Most
2. Almost
3. bad
4. badly
5. well

Exercise 4 Converting from Passive to Active Voice

1. The local construction firm will build 50 single-family homes.
2. The city council has proposed an increase in real estate taxes.
3. The builder announced a reduction in the price of the new homes.
4. The engineer submitted the plan for urban development to the town board.
5. Union Bank offers mortgage financing to first-time home buyers.
6. The landlord has submitted a plan to convert the building to cooperative ownership.
7. The Board of Directors vetoed renovations.
8. The bank rejected the mortgage application that the young couple submitted.
9. The speaker mentioned the difficulty in changing the existing building code.
10. The majority of the shareholders attended the annual meeting of the condominium.

CHAPTER 6: MAINTAINING HEALTHY COMPLEX SENTENCES

Activity A Building English Power: Introductory Clauses and Phrases and Parenthetical Expressions

1. For many workers, health care is a major expense.
2. If you wish to continue to participate in the insurance plan, please return the signed form.

3. To include vegetables and fruits in your daily diet is recommended for good health (**C**orrect)
4. Since your last bill, premium rates have increased slightly.
5. You will receive supplemental insurance benefits if you are hospitalized. (**C**orrect)
6. More than ever, you need adequate medical insurance.
7. We cannot process your claim, however, without a detailed statement.
8. In an effort to control costs, we increased the deductible on the dental plan.
9. We have served one million members since we offered our plan. (**C**orrect)
10. To exercise regularly is important for physical fitness. (**C**orrect)
11. As you know, a smoke-free workplace is endorsed by most Americans.
12. DWI laws, without a doubt, have become stricter in the past decade.
13. After the public debate, Congress will address the issue of health insurance.
14. Within the last few years, HMOs have become an alternative approach to medical treatment for patients.
15. Arthritis problems, according to some doctors, can be treated with a proper nutritional diet.

Activity B Building Word Power

1. ophthalmologist
2. cholesterol
3. nutrition
4. network
5. HMO
6. orthopedic
7. transition
8. Metabolism
9. nutritionist
10. comprehensive
11. Biomedical
12. outpatient
13. mammogram
14. DWI

Activity E Writing Practicum

Exercise 1 Replacing Descriptive Words *(Answers may vary.)*

1. The National Heart Association states that foods filled with saturated fats are disastrous (poor, detrimental) for your health.
2. Selecting a(n) comprehensive (quality, ideal) medical plan can be one of the most important choices a family can make.
3. Reading numerous health coverage brochures can be confusing (dull).
4. The nurse who administered my blood test was efficient (competent).
5. In addition to improving your health, a well-structured exercise plan can be invigorating (stimulating, exhilarating).
6. Filling out medical forms is a tedious (monotonous, boring) activity.
7. Everyone hopes that medical reimbursement checks are issued promptly (expeditiously, quickly).
8. The choice of a physician is a key (major) factor when selecting a health plan.
9. My yearly physical revealed that I was in excellent (good) health.
10. Recuperating after a surgical procedure can be a gradual process.

Exercise 2 Choosing Positive Opening Sentences in Letters of Adjustment and Refusal

A. *Topic: Refusing an Invitation*
Option: 1. __X__ 2. ______
Reason: The opening statement should recognize the importance of this organization's activity. Option 2 is curt and shows no interest or appreciation for the opportunity.

B. *Topic: Refusing an Invitation*
Option: 1. ______ 2. __X__
Reason: The buffer statement should express the writer's acknowledgment of the invitation in a positive manner and indicate the reason for nonacceptance. Option 1 is unacceptable because it starts with an upbeat remark but then turns negative without providing a reason for nonacceptance.

C. *Topic: Refusing a Claim*
Option: 1. ______ 2. __X__
Reason: Option 2 acknowledges the request and tells the reader what has to be done. Option 1 leaves the reader with insufficient information.

D. *Topic: Refusing a Claim*
Option: 1. __X__ 2. ______
Reason: Option 1 is empathetic and expresses the reason for declining the claim. Option 2 is cold.

E. *Topic: Refusing an Applicant*
Option: 1. __X__ 2. ______
Reason: Statement begins with a positive approach and shows empathy for the applicant. Option 2 conveys a negative attitude.

F. *Topic: Refusing an Applicant*
Option: 1. ______ 2. __X__
Reason: Option 2 acknowledges the client's interest and regrets not being able to offer the special package. Option 1 lacks interest in the customer's needs and displays poor goodwill.

G. *Topic: Requesting an Adjustment*
Option: 1. ______ 2. __X__
Reason: Option 2 gets to the point in a positive manner. Option 1 is somewhat arrogant.

H. *Topic: Requesting an Adjustment*
Option: 1. ______ 2. __X__
Reason: Option 2 uses a positive tone to imply that a billing error was made. Option 1 reflects an argumentative, accusatory tone.

CHAPTER 7: RETAILING COMMAS IN WORDS, PHRASES, AND CLAUSES

Activity A Building English Power: Comma Use in Series, Consecutive Adjectives, Nonrestrictive/Restrictive, Apposition, and Dates

1. You, a valued customer, are eligible for our money-saving bonus plan.
2. On May 1, 20xx, our popular, well-known rug sale will take place.

3. Price Cutter, a national chain, will open a store in Huntington, New York, a thriving suburban community.
4. Designer clothing, which appeals to today's shopper, can be purchased at discount stores.
5. Appliances carry a manufacturer's warranty that offers consumers protection against product defects. (**C**orrect)
6. Shoppers who look for affordable quality are joining warehouse price clubs. (**C**orrect)
7. Our electronics store, which is located in the mall, is offering a 10 percent discount on VCRs, camcorders, and CD players.
8. As a valued, respected charge customer, you are invited to our annual two-day sale, which will be held this weekend.
9. If you are interested in discount stores, Chet Edwards' latest guide, *The Popular Discount Malls,* is now available at your bookstore.
10. The seminar speaker discussed today's competitive market, consumer preferences, and new marketing trends.
11. The new, upscale mall that attracted huge crowds on the weekend will extend its shopping hours.
12. The popular home-furnishing store offers free shipping, a convenient layaway plan, and low credit rates.
13. The firm has scheduled Friday, July 31, for its annual physical inventory.
14. Our spring catalog, which was recently mailed to you, illustrates the latest women's fashions.
15. The best seller *Repackaging Your Sales Approach* sells for $19.95 plus shipping. (**C**orrect)

Activity B Building Word Power

1. warehouse clubs
2. boutiques
3. electronic malls
4. Megamalls
5. upsurge
6. redeemed
7. Innovative
8. backlogged
9. privileged
10. outlet malls
11. upscale
12. physical inventory
13. warranty
14. franchises

Activity E Writing Practicum

Parallelism

1. The goals of the manufacturer are to mass produce custom-made clothing and to sell directly to the consumer.
2. The antiquated marketing system has been updated by using computer technology and by sending employees for training.
3. In marketing, computers are used for product inventory control, daily sales records, and customer feedback.
4. Suburban malls have replaced downtown shopping areas because they are centralized, diversified, and accessible.
5. Our entire line of merchandise, which includes jewelry, furniture, and new fall fashions, is on sale.
6. This month's sales of winter clothes will be a huge success because of two-day sales and a 20 percent discount.
7. Come to the selling course prepared to take notes, to listen to the guest speaker, and to ask some questions.
8. Your mail-order catalog was received yesterday, and my order for furniture was sent to the purchasing department today.
9. Good salespersons should have knowledge of and familiarity with their stock.
10. The department store, specializing in sporting goods, also features designer clothing, exercise equipment, and travel books.
11. The mall has increased its budget for hiring security personnel and preventing shoplifting.
12. The company sent out direct mailers to advertise the sale and promote new merchandise.
13. When the shopping mall was designed, the architect considered mall motif, store layout, and accessibility to roads.
14. The shopper purchased groceries at the supermarket, cosmetics at the drug store, and a sweater at the boutique.
15. Service, dependability, and product value are the major goals of our firm.

CHAPTER 8: CHECK-IN WITH THE SEMICOLON

Activity A Building English Power: Commas and Semicolons

1. For hotel reservations, you can call your travel agent; or you can dial our toll-free number.
2. Hotels in major cities offer complimentary features in their weekend packages; for example, continental breakfast, health club, and free parking.
3. Popular vacation resort areas include San Juan, Puerto Rico; Cancun, Mexico; and Hilton Head, South Carolina.
4. A great way to visit new areas is to stay at bed-and-breakfast inns; they are informal and friendly.
5. The conference room has been reserved for our June meeting; however, plans for the luncheon are incomplete.
6. The convention activities will be held in Scottsdale, Arizona; in addition, local tours will be offered.
7. For your information, a layout of the Conference Center is attached; and a brochure about our facilities is enclosed.
8. Registration for the conference will take place in the hotel lobby; and the exhibit area, which is rather large, will be held on the mezzanine floor.
9. We, unfortunately, cannot accommodate your group during the week of May 15; but our facilities are available the following week.
10. The members of the Executive Board, after visiting several hotels, met to discuss future convention sites; but they were unable to reach a decision.
11. Reservations must be guaranteed with a credit card; otherwise, rooms are held only until 4 p.m.
12. Our hotel chain combines quality accommodations, fine service, and good value; moreover, these hotels are highly rated by travel agents.
13. A block of 20 rooms has been held for your company conference; you are reminded of our 48-hour cancellation policy.
14. We offer a special weekend rate; however, reservations must be made two weeks in advance.
15. Our hotels are located at the following airports: London, England; Milan, Italy; and Tokyo, Japan.

Activity B Building Word Power

1. consolidation
2. amenities
3. Gourmet
4. projections
5. implemented
6. complimentary
7. partition
8. directories
9. continental breakfast
10. concierge

Activity E Writing Practicum

Exercise 1 Misplaced Modifiers *(Answers may vary.)*

1. The hotel visitors riding the jitney saw the golf course.
2. Mr. Smith left his briefcase containing his wallet, his passport, and his important papers in the hotel room.
3. Ms. Garcia met a relative in the hotel lobby whom she hadn't seen since childhood.
4. The new hotel that was managed by Mr. Bright was built by Gutierrez Construction Co. in 1996 at a cost of $2 million.
5. The office clerk made a reservation for a room with double beds and a private bath for Mr. and Mrs. Robert Jiminez.
6. For the convenience of its guests, the Belle Convention Center has excellent facilities, good restaurants, and shuttle service to and from the airport.
7. Telephone reservations are accepted on a space-available basis for prospective guests.
8. For dinner, the dress code for men is formal with jackets.
9. Picture Waterfront Inn as a vacation spot with its landscaped beachfront and its rocky shoreline.
10. Business travelers use our deluxe hotels at two airports; namely, Chicago O'Hare and Los Angeles International.
11. The small hotel, a charming place, is within walking distance of the shopping district.
12. Spas are ideal for stressed individuals.
13. The concierge recommended to the guests a restaurant which is a short distance from the hotel.

14. The beach resort gave its guests a complimentary night.
15. The resort, nestled in the mountains, offers its guests a relaxed vacation.

Exercise 2 Dangling Modifiers

1. I saw a spectacular view of the countryside while riding in the hot-air balloon.
2. While she was traveling in the Southwest, her family was touring Europe.
3. The owner decided to sell the sculpture, which was hidden in the basement for the last 40 years.
4. The president arrived in town while I was a guest at the hotel.
5. After visiting the seaside village many times, I drew plans for a summer cottage.
6. After three days of rain, the visitors left the national park.
7. My assistant completed the report while I was getting ready for the staff meeting.
8. After my 10-hour flight, I was fatigued.
9. The people of New Zealand welcome visitors to see the breathtaking scenery.
10. Since we prefer the beaches to the mountains, the Caribbean was our choice.

Exercise 3 Modifier Errors *(Answers may vary.)*

(1) Anyone who has studied at the Acme School of Hotel Management can learn to make quick decisions in the field of hotel management in a matter of minutes. (2) Located in a thriving and bustling downtown metropolitan area, our campus offers students the benefits of internships at nearby hotels. Our workshops and seminars are led by well-trained, professional leaders in the hotel industry. Our brochure, which includes itemized costs and course descriptions, will be sent free of charge on request.

Exercise 4 Modifier Errors *(Answers may vary.)*

(1) Arriving at our hotel, we found that members of a Bird Watchers' Convention had been checked into our room. (2) Apologizing, the desk clerk led us to another room on the first floor. (3) The dining

room was crowded with people wearing binoculars, eating, and discussing the local migrating bird population. (4) There were other guests among the bird watchers, but they were hard to spot. (5) Rising early to spot the birds, the devotees woke us up each morning. (6) Eventually giving in, we were intrigued by the bird-watching excursions and decided to join in on the fun.

CHAPTER 9: MARKETING POSSESSIVES

Activity A Building English Power: Apostrophes

1. The firm**'s** decision to introduce video disks in its marketing plan will affect the advertising department.
2. Management met to discuss the use of telemarketing to increase the businesses**'** sales.
3. A multimedia approach provides opportunities to meet many readers**'** and listeners**'** needs.
4. David and Emily**'s** proposal to study the advantages of direct mailing was given to the marketing research group.
5. The applicant with five years**'** experience in fashion display was offered the department manager**'s** position.
6. Franchises offer an individual the opportunity to be one**'s** own boss.
7. The Jones**'s** and Burns**'s** companies are merging to increase their productivity.
8. Salespersons**'** attitudes often determine customers**'** decisions to purchase merchandise.
9. Competitors**'** prices are influencing the sales policies of the firm and its subsidiaries.
10. The women**'s** and men**'s** departments are being reorganized to include designer clothing.
11. We bought three similar CDs for our brothers-in-law**'s** birthdays.
12. The store is offering a 10 percent discount off the manufacturer**'s** suggested retail price.
13. Trucks are a popular means of transporting goods because merchandise can go directly from the shipper's place to the customer**'s** business.

14. The library**'s** new electronic information center will offer access to business periodicals.

15. Although Jared has the advantage of better locations for his shops than Eric, each of Jared**'s** and Eric**'s** repair shops has a good reputation for fast and satisfactory service.

Activity B Building Word Power

1. franchise
2. franchisee
3. commission
4. standalone
5. subsidiary
6. supplement
7. Telemarketing
8. Network marketing
9. entrepreneur
10. franchiser
11. feasibility
12. rebate

Activity E Writing Practicum (Answers may vary.)

Sentence Rewrite for Clarity

(1) Most people strongly emphasize personal health and appearance in their daily living. (2) However, very busy schedules don't always enable individuals to get needed nutrients.

(3) The Indigo Vitamin System is a nutritionally balanced combination of vitamins and minerals essential for a healthy body. (4) Furthermore, the Indigo Vitamin System encourages a weight management program that includes daily exercise, proper diet, and stress control.

(5) After following it for one year, I am so strongly committed to the Indigo Vitamin System that I have become an independent distributor of its nutritional line. (6) If you are impressed by the enclosed brochure and wish to order any of the exceptional products, please contact me. (7) You also can become an independent distributor after using the products for a trial period.

(8) As an independent distributor, you will represent an outstanding product line, which will give you a good income. (9) You will receive a discount on all items you purchase. (10) Whether you are a beginner or a veteran salesperson, you will receive the training support needed to provide your customers with superior service.

(11) Good health is the key to a successful and productive life.

(12) Accept our invitation to try the Indigo Vitamin System, which will give you a healthier, more active lifestyle.

CHAPTER 10: TRAVELING WITH COLONS AND PRONOUNS

Activity A Building English Power: Colons and Pronouns

Exercise 1 Who and Whom, That and Which

1. who
2. whom
3. who
4. whom
5. whom
6. Whomever, who
7. Whom
8. which
9. whom
10. whomever
11. whom
12. whom
13. who
14. who
15. who

Exercise 2 Who, Whom, Colon, and Other Punctuation

1. We plan to leave for the airport at 3:10 p.m. with whoever is available at that time.
2. Who is interested in touring the following cities: London, Milan, and Athens?
3. Tourists who are planning to travel by car should review the following checklist: get car tuneup, obtain roadmaps, get sightseeing information, and carry emergency supplies.
4. The manufacturing executives will visit their plants as follows: Bangkok, Thailand; Jakarta, Indonesia; and Kobe, Japan.
5. Who has scheduled the afternoon tour for 1:15 p.m.?

Activity B Building Word Power

1. emerging
2. itinerary
3. budget
4. cost conscious
5. nominal
6. segments
7. preliminary
8. priority
9. instantaneously
10. diversified
11. Globalization
12. liaison
13. accrues
14. strategies
15. no-frills
16. extended-stay hotels

Activity E Writing Practicum

Exercise 1 Pointing and Limiting Adjectives

1. Those kinds of package tours are inexpensive.
2. That tour guide was well informed.
3. Those cruises cater to senior citizens.
4. Please ask those travel agents to come to the tourism convention.
5. Are these types of flights eligible for frequent flier miles?
6. This brochure describes a very exciting trip to the Galapagos Islands.
7. I can't decide between these accommodations and those luxury rooms.
8. Those countries that are in the news because of terrorism show a sudden drop in tourism.
9. That bellhop took such good care of my luggage that I gave him a big tip.
10. These kinds of side trips are available for nominal fees after you pay for the total package.

Exercise 2 Writing Sentences for Conciseness and Clarity

1. People now prefer shorter vacations.
2. We decided to take an auto trip because of the increase in airfare.
3. If a strike occurs, the passengers will receive a refund.
4. Usually, individuals still prefer to discuss vacation plans with a travel agent.
5. At your request, we mailed you the travel brochure for Alaska.
6. Thank you for payment for your European vacation tour.
7. I am unable to attend next week's seminar.
8. Thank you for your support and help.
9. Although we have sent you numerous reminders, we have not received your check for the overdue balance of $500.
10. We are sending you our check for $100 to reserve the meeting room.

Exercise 3 Writing Paragraphs for Conciseness and Clarity (Answers may vary.)

Paragraph 1

Are you planning a trip this summer? A fast, efficient way to do this is to use Trip Saver, a computer software package to help you plan your vacation. This package will map out a route, suggest side trips, and give you estimated driving time. This program also will list overnight accommodations. For a quick, scenic way to travel, use Trip Saver.

Paragraph 2

Join our Five Star Travel Club to get a low-priced vacation of $199 for each member of your family. Members receive reduced rates on airfare, hotel accommodations, and car rentals. You may travel to exotic Hawaiian resorts or to cosmopolitan U.S. cities.

To be eligible for these inexpensive vacations, join our club today. Mail the enclosed card; we will bill you for the $25 membership fee.

CHAPTER 11: GOVERNING QUOTATIONS

Activity A Building English Power: Quotation Marks

1. The town supervisor stated, "If you have any suggestions on ways in which we can improve the town, please contact my office."
2. "Property taxes will increase by 10 percent," stated the controller, "unless we reduce spending."
3. "Are you interested in the newly formed civic association?" asked my neighbor.
4. The judge suggested that a community service program could be an alternative means of punishment for minor offenses. (reported speech)
5. The notice read as follows: "A public hearing to discuss housing will be held on April 15, 2005, at Town Hall."
6. "You, an active civic leader, are invited to attend the meeting on Thursday," stated the council president.
7. I asked, "Have you seen the article titled 'Federal Budget' that appeared in the *Tribune Herald* last week?"
8. Did the community leader say, "We need to have greater recreation facilities for our youth"?

9. Do you plan to attend the program called "Multicultural Interaction" that is being sponsored by the Community Development Department? (reported speech)
10. "As you know," commented the congresswoman, "we are introducing legislation to give financial assistance to college students."
11. My friend asked, "Have you read the article 'The Role of Government in the 1990s'?"
12. "With the changes in traffic patterns," announced the legislator, "we need to revise our noise ordinance."
13. "Is it possible that the trial will be postponed?" asked the lawyer.
14. In honoring the town resident Jim O'Connor, the mayor said, "Jim's dedication to his neighbors has helped our quality of life."
15. The civic association member asked, "Why were the beautification plans for Main Street postponed until the spring?"

Activity B Building Word Power

1. fiscal
2. multicultural
3. legislation
4. stalled
5. diminished
6. budget
7. legislator
8. town council
9. municipal
10. alternative
11. bulk
12. ordinance
13. commissioner
14. assessment

Activity E Writing Practicum

Exercise 1 Using Prepositions Correctly

1. of
2. between
3. with
4. to
5. from
6. in
7. to
8. on
9. for
10. about
11. about
12. among
13. by
14. at
15. during

Exercise 2 Puzzling Words

1. fewer
2. less
3. may
4. can
5. set
6. sit
7. lose
8. loose
9. allot
10. between
11. among
12. Sometime
13. some time
14. They're
15. their
16. Anyone
17. Any one
18. everyone
19. Every one
20. Regardless
21. lay
22. lie

CHAPTER 12: HELP WANTED APPLYING HYPHENS AND NUMBERS CORRECTLY

Activity A Building English Power Hyphens and Number Usage

1. Employers are interested in self-motivated individuals who are well informed about the firm'**s** products.
2. The regular conference fee is **$20;** members have a 10 **percent** discount.
3. Applicants should send a follow-up letter to thank the interviewer.
4. The well-known speaker Dr. Robert Reed will present a two-day workshop on self-assertiveness.
5. The applicant's interview was scheduled for **2** p.m. at the company's headquarters at **One** Park Row.
6. The firm budgeted **$1.2 million** for renovation of its 50-story building.
7. At the conference, the vice president of the firm remarked that the sales force will be reduced by one-third by mid-April.
8. The new contract includes a three-week vacation for employees who have been in the company for **five** years.

9. Home-based work is a good option for part-time employment.
10. The 12-page manual will give you detailed, step-by-step instructions for the new computer software.
11. Seventeen- and eighteen-year-olds are encouraged to find volunteer internships to gain hands-on experience.
12. Entry-level jobs have become more demanding with the need for higher technical skills.
13. The president of the company is well known in the electric utility industry.
14. Some individuals prefer the freedom of a freelance job rather than a traditional 8-hour position.
15. <u>**Twenty**</u> applicants were interviewed for the <u>**three**</u> positions as paralegals.

Activity B Building Word Power

1. modulated
2. teleconferencing
3. human resources
4. commitment
5. assertiveness
6. assess
7. internship
8. preliminary
9. crucial
10. prospective
11. perusal

Activity E Writing Practicum

Revisions

1. A (correct format)
2. C (clear explanation of job source)
3. B (grammatically correct)
4. B (clear and complete statement)
5. B (a comprehensive and well-written statement)
6. C (complete, punctuated correctly)

7. B (correct grammar and number usage)
8. D (concise and correctly written)
9. A (correct format)
10. C

Rekeyed Letter

Dear Mr. Shane:

I am writing to you at the suggestion of my neighbor, Mrs. Dorothy Frank, one of your employees. She informed me that there are five summer internships available in your office, and I am most interested in applying for one of them. At Queens College, I am majoring in municipal government, a career field that I find fascinating.

Please send an internship application form to me at the above address and include any other requirements needed in applying for the internship. If necessary, I can provide three letters of reference. I am enclosing a self-addressed, stamped envelope for your convenience.

I am available any day after 2 p.m. for an interview. I look forward to hearing from you.

Sincerely yours,

Michelle Yager

Enclosure

Appendix B

Model Documents

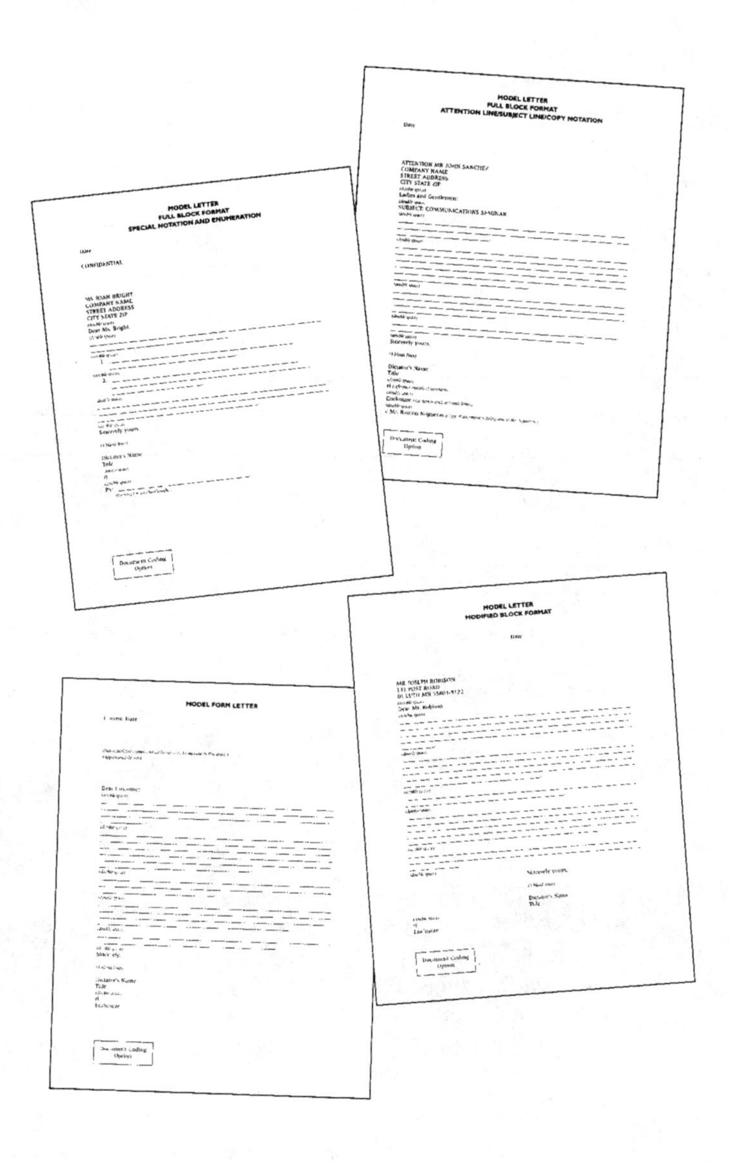

Model Letter
Full Block Format
Attention Line/Subject Line/Copy Notation

Date

ATTENTION MR JOHN SANCHEZ
COMPANY NAME
STREET ADDRESS
CITY STATE ZIP
(double space)
Ladies and Gentlemen:
(double space)
SUBJECT: COMMUNICATIONS SEMINAR
(double space)

____
____
______________________.

(double space)

____
____
____.

(double space)

____
______________________.

(double space)

____
______________________.

(double space)
Sincerely yours,

(3 blank lines)
Dictator's Name
Title
(double space)
ri *(reference initials of secretary)*
(double space)
Enclosure *(An item is enclosed with letter.)*
(double space)
c Mr. Ramon Nogueras *(Copy of document is being sent to Mr. Nogueras.)*

Document Coding Option

Model Letter
Full Block Format
Special Notations and Enumeration

Date

CONFIDENTIAL

MS JOAN BRIGHT
VICE PRESIDENT
COMPANY NAME
STREET ADDRESS
CITY STATE ZIP
(double space)
Dear Ms. Bright:
(double space)

____

____
(double space)

1. ________________________________

   __________________.

(double space)

2. ________________________________

   ____

   __________________.

(double space)

____

____.
(double space)
Sincerely yours,

(3 blank lines)

Dictator's Name
Title
(double space)
ri
(double space)
PS: __________________.
(Postcript is an afterthought.)

Document Coding Option

Model Letter
Modified Block Format

Date

MR JOSEPH ROBISON
131 POST ROAD
DULUTH MN 55803-5122
(double space)
Dear Mr. Robison:
(double space)

____
____
_________________________.
(double space)
____
____
_________________________.
(double space)
____
____________________.
(double space)
____
____
_________________________.
(double space)
____
____
_________________________.
___________.
(double space)

Sincerely yours,

(3 blank lines)

Dictator's Name
Title

(double space)
ri

Enclosure

Document Coding Option

Model Form Letter

Current Date

(Individualized names and addresses may be merged in this space.)
(Approximately 10x)

Dear Customer:
(*double space*)

____
____
______________________________.
(*double space*)

____
____
____.
(*double space*)

____
____
______________________________.
(*double space*)

____
____
______________________________.
(*double space*)
Sincerely,

(*3 blank lines*)

Dictator's Name
Title
(*double space*)
ri

Enclosure

Document Coding Option

Large (No. 10) Envelope

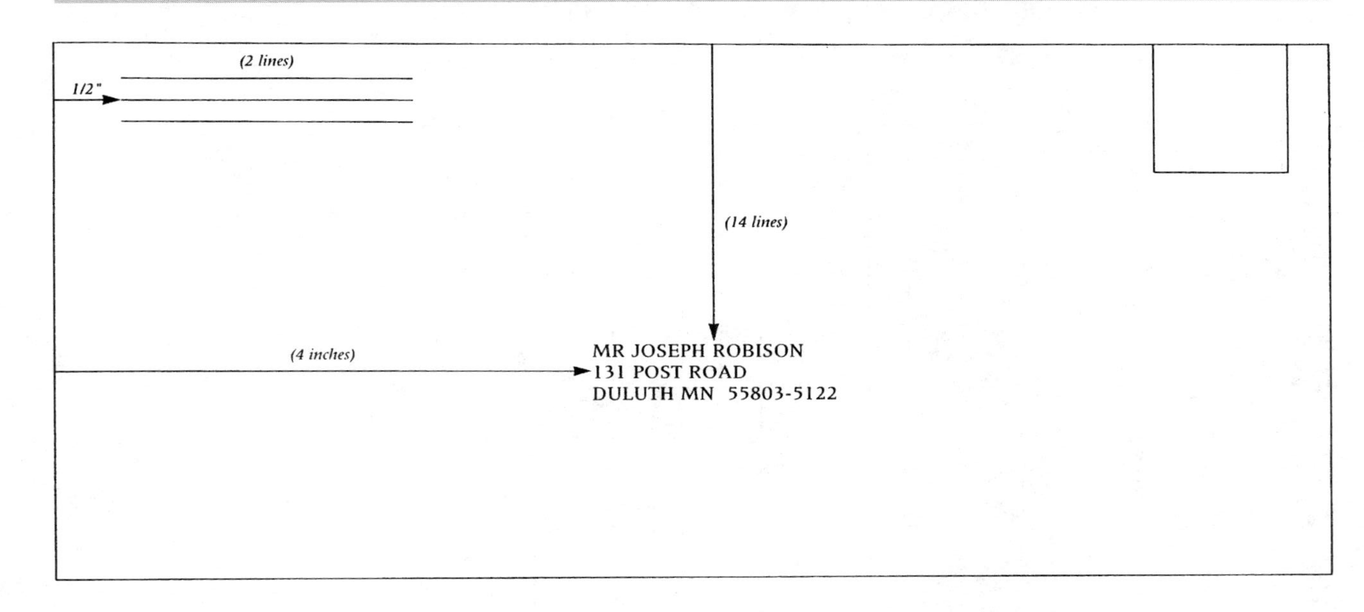

(shows placement of **confidential** and **registered** notations on both large and small envelopes)

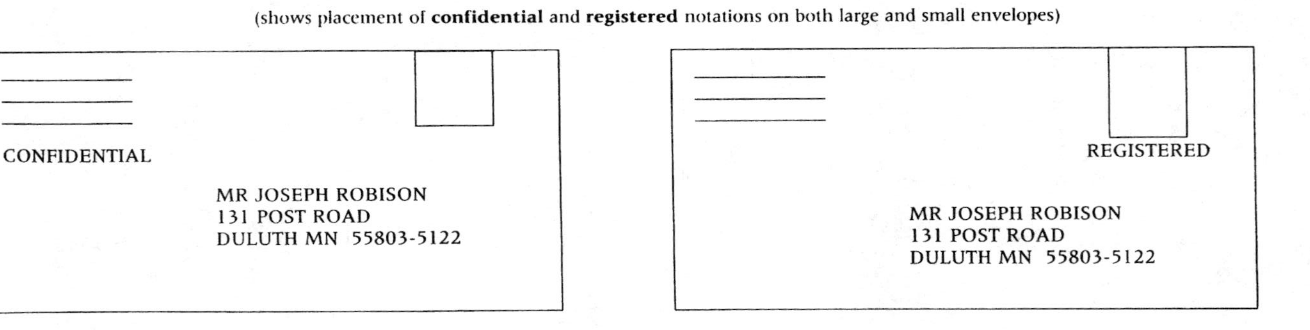

Small (No. 6 3/4) Envelope

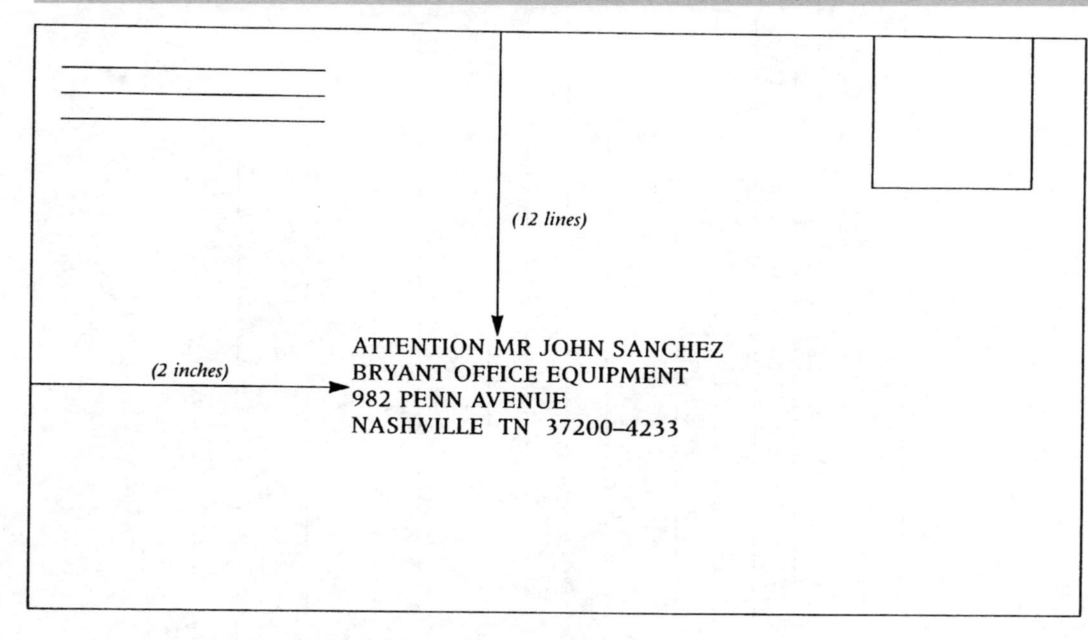

Interoffice Memorandum Format

TO: ____________________
(double space)
FROM: ____________________
(double space)
DATE: ____________________
(double space)
SUBJECT: ____________________
(triple space)

____
___________________________________.
(double space)
____
____
____
____.
(double space)
____
____
____
____.
(double space)

Enumerated items using bullets

- ____
  __________________________________.

(double space)

- ____.

(double space)

- ____
  __________________________________.

(double space)
ri

Enclosure
(double space)
c Carla Green, Director

Document Coding Option

E-Mail Correspondence

Subject of E-mail

File Edit View Tools Message Help

Reply Reply All Forward Print Delete Previous Next

Receiver's e-mail address

To: cruz@aol.com
Cc:

Purpose of message

Subject: Subscription Renewal
Attachments: none

Salutation included; message sent outside company

Dear Ms. Cruz:

We acknowledge your subscription renewal to Money Matters. Every Friday online you will receive a weekly update of financial news.

Message is concise and clear

Thank you for renewing.

Contact information

Laura Crawford
Subscription Manager
lcrawford@aol.com

E-Mail Memorandums

Subject of E-mail

File Edit View Tools Message Help

Reply Reply All Forward Print Delete Previous Next

To: O'Brien.sue@packards.com
Cc:
Subject: Meeting for Sales Staff
Attachments: none

Informal style

Just a reminder that Wednesday's meeting to discuss the sales promotional campaign for the Notethink laptop is scheduled for 10 a.m. See you then.

Salutation and closing optional

Model Report

(2 inches)

TITLE
(triple space)

____

____

______________________________.

(triple space)

SIDE HEADING
(double space)

____

____

______________________________.

(triple space)

SIDE HEADING
(double space)

____

____

______________________________.

(triple space)

SIDE HEADING
(double space)

____

____

______________________________.

(Note: Spacing may vary when keying on computer. If keyed as a single-spaced business report, do not indent first line of paragraph.)

Appendix C

Addressing Government Officials

President of the United States

THE PRESIDENT
THE WHITE HOUSE
WASHINGTON DC 20500-0000

Dear Mr. President:

United States Senator

THE HONORABLE (FULL NAME)
UNITED STATES SENATE
WASHINGTON DC 20510-0000

Dear Senator (Surname):

Governor

THE HONORABLE (FULL NAME)
GOVERNOR OF (STATE)
CAPITAL CITY STATE ZIP CODE

Dear Governor (Surname):

Member of State Legislature

THE HONORABLE (FULL NAME)
THE STATE SENATE OR THE HOUSE
OF REPRESENTATIVES
CAPITAL CITY STATE ZIP

Dear Mr./Ms. (Surname):

United States Congressman/
Congresswoman

THE HONORABLE (FULL NAME)
HOUSE OF REPRESENTATIVES
WASHINGTON DC 20515-0000

Dear Mr./Mrs. (Surname):

Mayor

THE HONORABLE (FULL NAME)
MAYOR OF (CITY)
CITY STATE ZIP CODE

Dear Mayor (Surname):

Appendix D

Postal Abbreviations

State	Abbreviation	State	Abbreviation
Alabama	AL	Nebraska	NE
Alaska	AK	Nevada	NV
Arizona	AZ	New Hampshire	NH
Arkansas	AR	New Jersey	NJ
California	CA	New Mexico	NM
Colorado	CO	New York	NY
Connecticut	CT	North Carolina	NC
Delaware	DE	North Dakota	ND
District of Columbia	DC	Ohio	OH
Florida	FL	Oklahoma	OK
Georgia	GA	Oregon	OR
Hawaii	HI	Pennsylvania	PA
Idaho	ID	Puerto Rico	PR
Illinois	IL	Rhode Island	RI
Indiana	IN	South Carolina	SC
Iowa	IA	South Dakota	SD
Kansas	KS	Tennessee	TN
Kentucky	KY	Utah	UT
Louisiana	LA	Vermont	VT
Maine	ME	Virgin Islands	VI
Maryland	MD	Virginia	VA
Massachusetts	MA	Washington	WA
Michigan	MI	West Virginia	WV
Minnesota	MN	Wisconsin	WI
Mississippi	MS	Wyoming	WY
Missouri	MO		
Montana	MT		

Appendix E

Proofreaders' Marks

Mark	Meaning	Draft Copy	Corrected Copy
	delete	I will not attend.	I will attend.
^	insert	inspect for breakage (and test)	inspect and test for breakage
STET	leave it in	role of key persons	role of key persons
¶	new paragraph	The check cleared. ¶Your statement	The check cleared. Your statement
	transpose	sturdy and durable equimpent	durable and sturdy equipment
	move copy as shown	well known. The speaker was	The speaker was well known.
≡	capitalize	The World of Bytes	THE WORLD OF BYTES
/	make lowercase	The Local branch	The local branch
#	space	Hemay be	He may be
○	spell out	(5) CDs	five CDs
‖	align copy	TO: John Woods FROM: Rosa Bell	TO: John Woods FROM: Rosa Bell
	close up	how ever	however
(:)	insert colon	the list below pens paper clips, and disks	the list below: pens, paper clips, and disks
^,	insert comma		
(.)	insert period	I saw Mr Smith.	I saw Mr. Smith.

Appendix F

Personal Dictionary

The words in this Appendix appear in Activity B, "Building Word Power," in each chapter.

accelerate (*v.*; 2)—to progress faster

accrues (*v.*; 10)—to be added or increased periodically

alternative (*n.*; 2, 11)—choice between two or more things

amenities (*n.*; 8)—extra services offered to customers

ASAP (adv.; 9)—abbreviation for "as soon as possible"

assertiveness (*n.*; 12)—the quality of being self-confident or forceful

assess (*v.*; 12)—to evaluate

assessment (*n.*; 11)—an official evaluation of taxable property

ATM (*n.*; 1)—automated teller machine

authorization (*n.*; 1)—legal power to complete a transaction for another person

backlogged (*adj.*; 7)—refers to tasks or materials accumulated and unperformed

biomedical (*adj.*; 6)—term describing medical science that deals with the relationship between body chemistry and function

blue chip stocks (*n.*; 3)—stocks of corporations that are leaders in their fields and have demonstrated consistent growth

bodily injury insurance (*n.*; 2)—physical damage coverage

bonds (*n.*; 3)—interest-bearing certificates issued by government or business to raise money

boutique (*n.*; 7)—small retail shop selling current, trendy clothing

budget (*n.*; 10, 11)—plan of income and projected expenses

bulk (*n.*; 11)—large quantity

cafeteria-type medical insurance (*n.*; 2)—type of insurance for which the individual chooses portions of coverage related to his or her needs

catastrophic insurance (*n.*; 2)—insurance covering expenses from a major accident or disaster not covered by other insurance

certificate of deposit (CD) (*n.*; 1)—savings account for a specific period of time that offers a higher rate of interest and carries a penalty for early withdrawal

cholesterol (*n.*; 6)—a fatty substance found in the body

co-applicant (*n.*; 1)—a person who files a joint request in which both individuals are financially responsible

collateral (*n.*; 1)—item of value, such as stocks or property, used to secure a loan

commission (*n.*; 9)—additional pay added to a salary based on total sales

commissioner (*n.*; 11)—individual who is the head of a department or division of a governmental agency

commitment (*n.*; 12)—pledge to do something, dedication

complimentary (*adj.*; 8)—free of charge

comprehensive (*adj.*; 6)—inclusive; of broad scope

concierge (*n.*; 8)—an individual who arranges special services for hotel guests

condominium (*n.*; 5)—building complex in which units are individually owned

configuration (*n.*; 4)—arrangement of parts making up the whole

consolidation (*n.*; 8)—grouping or merging together

continental breakfast (*n.*; 8)—breakfast consisting of a beverage and bread or light pastry

cooperative (*n.*; 5)—building complex in which owners of an apartment are shareholders of the building

corporate park (*n.*; 5)—cluster of office buildings in a park-like setting

cost conscious (*adj.*; 10)—sensitivity to costs of products or services

crucial (*adj.*; 12)—important; significant

debit card (*n.*; 1)—card used to authorize transactions deducted directly from the checking account

debris (*n.*; 5)—rubbish or the remains of something destroyed

deductible (*n.*; 2)—amount of insured's liability for a claim

delinquent (*adj.*; 5)—business term describing a past-due account

diminished (*v.*; 11)—reduced, lessened

directories (*n.*; 8)—lists of names, places, and items

diversified (*adj.*; 10)—varied

DWI (*n.*; 6)—acronym for "driving while intoxicated"

electronic banking (*n.*; 1)—banking transactions completed through electronic machines

electronic mail (e-mail) (*n.*; 4)—system through which messages are transmitted via computer networks

electronic malls (*n.*; 7)—system on the Internet through which merchandise can be purchased via the computer

emerging (*v.*; 10)—evolving into something new or improved

entrepreneur (*n.*; 9)—individual who initiates, organizes, and manages a business

escalate (*v.*; 2)—to grow or increase

excess (*n.*; 1)—extra

extended-stay hotels (*n.*; 10)—accommodations that offer homelike suites with kitchen facilities; rates are based on longer stays

facsimile equipment (fax) (*n.*; 4)—telephone and electronic equipment through which an exact copy of information is transmitted

FDIC (*n.*; 1)—Federal Deposit Insurance Corporation, a federal agency that insures savings deposits up to $100,000

feasibility (*n.*; 9)—capability or practicality of completing an activity

fiscal (*adj.*; 11)—pertaining to financial matters

franchise (*n.*; 7, 9)—licensing right to run a business or service for a well-known company according to set operating procedures; e.g., McDonald's™ franchise

franchisee (*n.*; 9)—individual or firm that has been granted a franchise

franchiser (*n.*; 9)—company or manufacturer that grants licensing rights to individuals to run a business

global (*adj.*; 1)—worldwide

globalization (*n.*; 10)—the act of becoming or making worldwide

glossary (*n.*; 3)—list of terms in a special subject with definitions

gourmet (*adj.*; 8)—term describing specialized, fancy foods

HMO (*n.*; 6)—health maintenance organization; group offering medical services to its subscribers

human resources (*adj.*; 12)—term used to identify personnel departments in companies

implemented (*v.*; 8)—put into effect

index (*n.*; 5)—guide used to measure items such as prices or wages

innovative (*adj.*; 7)—term used to describe new ideas or methods

instantaneously (*adv.*; 10)—immediately

insufficient funds (*n.*; 1)—not enough money in an account to clear checks for payment

Internet (*n.*; 4)—system that allows users to communicate with others throughout the world via computers

internship (*n.*; 12)—voluntary work assignment without pay

investment (*n.*; 3)—the use of money for the purpose of receiving profit

investment portfolio (*n.*; 3)—list of stocks and bonds owned by an individual, bank, or business

IRA (*n.*; 1)—individual retirement account; a savings plan that allows individuals to put aside money that is not taxed until retirement

itinerary (*n.*; 10)—schedule of travel dates, accommodations, and events

laptop computer (*n.*; 4)—compact, lightweight computer that is easy to carry

lease (*n.*; 5)—contract between landlord and tenant specifying terms of agreement (e.g., length of contract, monetary deposit on apartment)

legislation (*n.*; 11)—laws proposed by local, state, and federal government

legislator (*n.*; 11)—individual elected or appointed to a law-making body

liability insurance (*n.*; 2)—protection against injury and property damages of another person

liaison (*n.*; 10)—a person who acts as a connection between individuals or organizations

mammogram (*n.*; 6)—x-ray examination to detect breast cancer

megamall (*n.*; 7)—large building or site in which retail stores, restaurants, entertainment, and accommodations are located

metabolism (*n.*; 6)—process by which the body converts food and oxygen into energy

modem (*n.*; 4)—electronic device used to convert and transmit data to or from a computer over telephone lines

modulate (*v.*; 12)—to regulate

monitor (*n.*; 4)—a computer screen

mortgage (*n.*; 1, 5)—a loan given by a bank to finance the cost of property, which is security for repayment of the loan

mouse (*n.*; 4)—small, hand-held device that moves the cursor on the computer screen

multicultural (*adj.*; 11)—of diversified ethnic backgrounds

multinational (*adj.*; 3)—referring to many nations

municipal (*adj.*; 11)—pertaining to the local government of a town or city

mutual fund (*n.*; 3)—group formed to invest money obtained from shareholders; investments include stocks and bonds

network (*n.*; 1, 6)—system of interconnecting equipment used for communications; group of individuals who cooperate on a project

network marketing (*n.*; 9)—selling approach in which individuals use personal contacts to make sales

networking (*n.*; *adj.*; 4)—systems of computers linked together to transmit data

no-frills (*adj.*; 10)—no extra amenities; basic services

nominal (*adj.*; 10)—very small amount in proportion to the value of the whole; for example, *nominal* fee for shipping and handling

notebook (*n.*; 4)—a portable, lightweight computer

nutrition (*n.*; 6)—nourishment that promotes good health

nutritionist (*n.*; 6)—individual specializing in nutrition

ophthalmologist (*n.*; 6)—medical doctor specializing in functions and diseases of the eyes

option (*n.*; 2)—right to choose

ordinance (*n.*; 11)—local law or rule

orthopedic (*adj.*; 6)—dealing with treatment, disease, and injuries of the spine, bones, joints, and muscles

outlet mall (*n.*; 7)—mall with discount retail stores

outpatient (*n.*; 6)—person receiving medical treatment or care but not requiring an overnight hospital stay

partition (*n.*; 8)—wall separating two areas

periodically (*adv.*; 5)—recurring from time to time

perusal (*n.*; 12)—act of reading or examining

phasing out (*v.*; 2)—gradually bringing to an end

physical inventory (*n.*; 7)—manual handcount of merchandise

potpourri (*n.*; 8)—combination, collection, mixed

preliminary (*adj.*; 10, 12)—introductory

premium (*n.*; 2)—the amount paid for an insurance policy

principal (*n.*; 5)—when used as a financial term, refers to the amount of money owed, not including interest

priority (*n.*; 10)—importance

privileged (*adj.*; 7)—entitled to a special advantage

projections (*n.*; 8)—studies predicting future trends

property damage insurance (*n.*; 2)—insurance against claims if the insured's car damages someone's property or land

prorate (*v.*; 5)—to divide proportionately

prospective (*adj.*; 12)—potential, expect to happen

prospectus (*n.*; 3)—report reviewing the financial condition, terms of ownership, and organizational plan of a new business

protocol (*n.*; 4)—set of standards or procedures that allow computers to interact

proxy (*n.*; 3)—stockholders who cannot attend an annual meeting permit another person to vote on issues on their behalf

realtor (*n.*; 5)—individual who is licensed by the National Association of Real Estate Boards to deal in real estate

rebate (*n.*; 9)—deduction off the original price of an item

redeem (*v.*; 7)—to pay off or buy back

reference (*n.*; 5)—person who can offer personal or employment information on an individual; recommendation

securities (*n.*; 3)—stocks and bonds

security deposit (*n.*; 5)—sum of money given as protection against losses resulting from damages or failure to pay rent

segment (*n.*; 10)—portion or section

software (*n.*; 4)—programs used on a computer (e.g., Microsoft® *Word, Excel*)

specialist (*n.*; 1)—individual devoted to a particular field of study

stalled (*v.*; 11)—delayed or put off

standalone (*adj.*; 9)—term used to describe a computer that operates independently of a network system

stockbroker (*n.*; 3)—agent who buys and sells stocks and bonds for a client

stocks (*n.*; 3)—certificates that show ownership in a corporation

strategies (*n.*; 10)—skills or plans used to reach a goal

subsidiary (*adj.*; 9)—providing secondary or support service to the main organization

subsidized (*v.*; 2)—supported with financial help

supplement (*v.*; 9)—to add to

telecommunications (*n.*; 4)—communications using telephone and video equipment

teleconferencing (*n.*; 12)—use of telephone and video equipment to conduct a joint meeting from two or more office sites

telemarketing (*n.*; 9)—use of the telephone to sell goods or services

town code (*n.*; 5)—body of laws passed by town regulating various procedures (e.g., building, environmental, and fire codes)

town council (*n.*; 11)—group of elected officials who run local government

transaction (*n.*; 1)—business deal or agreement

transition (*n.*; 6)—change from one stage of activity to another

umbrella insurance (*n.*; 2)—type of insurance that protects the insured against losses not covered by regular insurance

upscale (*adj.*; 7)—above the average

upsurge (*n.*; 7)—upward movement or increase

videoconferencing (*n.*; 4)—system that uses computer terminals and communications equipment to display data and images

warehouse club (*n.*; 7)—large retail store selling items in large quantities at discounted prices for its members

warranty (*n.*; 7)—guarantee by manufacturer that quality and service of product are satisfactory

web site (*n.*; 3)—a group of "pages" on the Internet made available by a business, organization, or individual that gives specific information

Index

C

D

E

Q

R